Philosophy of Time and Perceptual Experience

This book explores the important, yet neglected, relationship between the philosophy of time and the temporal structure of perceptual experience. It examines how time structures perceptual experience and, through that structuring, the ways in which time makes perceptual experience trustworthy or erroneous.

Sean Enda Power argues that our understanding of time can determine our understanding of perceptual experience in relation to perceptual structure and perceptual error. He examines the general conditions under which an experience may be sorted into different kinds of error such as illusions, hallucinations, and anosognosia. Power also argues that some theories of time are better than others at giving an account of the structure and errors of perceptual experience. He makes the case that tenseless theory and eternalism more closely correspond to experience than tense theory and presentism. Finally, the book includes a discussion of the perceptual experience of space and how tenseless theory and eternalism can better support the problematic theory of naïve realism.

Philosophy of Time and Perceptual Experience originally illustrates how the metaphysics of time can be usefully applied to thinking about experience in general. It will appeal to those interested in the philosophy of time and debates about the trustworthiness of experience.

Sean Enda Power is a research affiliate of University College Cork (UCC), a former visiting fellow at the Centre for Time, Sydney, and a former Irish Research Council Postdoctoral Fellow. His publications include 'Complex Experience, Relativity and Abandoning Simultaneity' (*Journal of Consciousness Studies* 2010), 'The Metaphysics of the "Specious" Present' (*Erkenntnis* 2013), and 'Perceiving Multiple Locations in Time: A Phenomenological Defence of Tenseless Theory' (*Topoi* 2015).

Routledge Studies in Metaphysics

Freedom of the Will
A Conditional Analysis
Ferenc Huoranski

The Future of the Philosophy of Time
Edited by Adrian Bardon

Properties, Powers and Structures
Issues in the Metaphysics of Realism
Edited by Alexander Bird, Brian Ellis, and Howard Sankey

The Puzzle of Existence
Why Is There Something Rather than Nothing?
Edited by Tyron Goldschmidt

Neo-Davidsonian Metaphysics
From the True to the Good
Samuel C. Wheeler III

Neo-Aristotelian Perspectives in Metaphysics
Edited by Daniel D. Novotný and Lukáš Novák

Nominalism about Properties
New Essays
Edited by Ghislain Guigon and Gonzalo Rodriguez-Pereyra

Substance and the Fundamentality of the Familiar
A Neo-Aristotelian Mereology
Ross D. Inman

Philosophy of Time and Perceptual Experience
Sean Enda Power

For a full list of titles in this series, please visit www.routledge.com

Philosophy of Time and Perceptual Experience

Sean Enda Power

Routledge
Taylor & Francis Group
New York London

First published 2018
by Routledge
52 Vanderbilt Avenue, New York, NY 10017

and by Routledge
2 Park Square, Milton Park, Abingdon, Oxon OX14 4RN

First issued in paperback 2020

Routledge is an imprint of the Taylor & Francis Group, an informa business

Library of Congress Cataloging-in-Publication Data
Names: Power, Sean Enda, author.
Title: Philosophy of time and perceptual experience / by Sean Enda Power.
Description: 1 [edition]. | New York : Routledge, 2018. | Series: Routledge studies in metaphysics ; 11 | Includes bibliographical references and index.
Identifiers: LCCN 2017061299 | ISBN 9780815370383 (hardback : alk. paper)
Subjects: LCSH: Time. | Perception. | Experience.
Classification: LCC BD638 .P69 2018 | DDC 115—dc23
LC record available at https://lccn.loc.gov/2017061299

ISBN 13: 978-0-367-66708-5 (pbk)
ISBN 13: 978-0-8153-7038-3 (hbk)

Typeset in Sabon
by Apex CoVantage, LLC

Contents

Preface vi
Acknowledgements viii

1 Introduction 1

2 The Philosophy of Time 21

3 The Structure of Perceptual Experience 39

4 Erroneous Experience 72

5 Choosing Erroneous Experience 99

6 Spatial Relations to the Past 112

7 The Perceptual Experience of Depth 138

8 Distortions of Depth 151

9 Temporal Experience 165

10 Eliminating Hallucination 202

11 Conclusion 227

Bibliography 239
Index 247

Preface

This book argues for an important, yet neglected, relationship between time and perceptual experience. This is the relationship between the philosophy of time and the temporal structure of perceptual experience. It examines how time structures perceptual experience and, through that structuring, how time makes perceptual experience trustworthy or erroneous. I argue that perceptual experience's temporal structure (and, as a result, the trustworthiness of particular perceptual experiences) depends on which philosophical theory of time is true.

Much of the literature on the relationship between time and experience considers time only as a kind of *content* of experience. It takes questions about time and experience as falling under questions such as how we experience time, how time appears to us, and how experience represents time. When applied to philosophical debates about time, common discussions turn on the relationship between such experiential content and philosophical claims about time. For example, some philosophers claim that temporal passage is a metaphysical feature of the world. A common question arising from experience is whether there is a phenomenology of temporal passage and, if there is, what the phenomenology means for the philosophical debate about passage.

However, this relationship between experiential content and the metaphysics of time is not the only relationship between them. There is another possible relationship: Philosophical theories of time can also influence theories of the real structure of experience itself. Furthermore, through the resulting theories of experience's structure, theories of time can influence interpretations of particular perceptual experiences as erroneous.

This relationship between time and experience is also an important one for experience. Indeed, it is arguably more important than that found in temporal content. For an understanding of the possible temporal structure of experience determines how one can answer questions about the content of experience in general, including its classification as veridical or erroneous.

In this book, I explore this latter relationship between perceptual experience and time in two ways.

First, as said, I argue that our understanding of time can determine our understanding of perceptual experience in relation to perceptual structure and perceptual error. In doing so, the book also examines the general conditions under which an experience may be sorted into different kinds of error—e.g., illusions, hallucinations, and anosognosia, a kind first explicitly described here (for reasons that are explained).

Second, I argue that some theories of time are better than others are at giving an account of the structure and errors of perceptual experience. The theories of tenseless theory and eternalism are better than their opposing positions of tense theory and presentism. Interpretations of experience in terms of tenseless theory and eternalism frequently allow veridical experiences where interpretations in terms of tense theory or presentism require erroneous experiences. Furthermore, even in cases in which the former require erroneous experiences, the errors are typically less severe than the errors to which one is committed through the latter.

This is an unexpected result. It is common to think that experience corresponds more closely to presentism and tense theory over eternalism and tenseless theory. I argue that, with respect to perceptual experience, once we go beyond time as experiential content (temporal experience) and move on to time as something that structures experience, then the opposite is the case.

Finally, throughout this book, I illustrate the effect on thinking about experience from thinking about time by considering the temporal structure of another important kind of experience: the perceptual experience of space. I argue that if we analyse the perceptual experience of space under the tenseless theory and eternalism, then we can better support an intuitive but problematic theory of perception and naïve realism than if we analyse it in terms of tense theory and presentism.

As such, this book is not just a defence of one theory of time in the face of others. It is an illustration of how we can usefully apply theories of time to thinking about experience in general. This book is an argument for the relevance of the philosophy of time to wider debates about the trustworthiness of experience. As briefly sketched at the end, experience may be merely the first in many areas it is relevant.

Acknowledgements

The ideas in this book developed over several years, with parts of this work going back to my time as an Irish Research Council (IRC) Postdoctoral Fellow in 2012 and other parts going even further to ideas arising during my PhD studies. The decision to write this book came much later during my postdoctoral career, inspired by discussions with and work by several friends and peers, from many different backgrounds and perspectives (academic and non-academic). As such, many people have informed the content of this book, far more than can be acknowledged here. I am very grateful to everyone who has helped me produce this work, whether by their thoughtful engagement with me in conversation, through moral support and encouragement, or through ruthless and serious criticism. Some of those who helped me require particular thanks.

I am indebted to Adrian Bardon and Giuliano Torrengo, who separately provided serious and thorough comments on the penultimate drafts. In addition, I am grateful to those who have discussed these ideas in depth with me over the years, especially Valtteri Arstila, Lilian O'Brien, Joel Walmsley, Robin Le Poidevin, Marc Wittmann, Virginie Van Wassenhove, and Julia Jansen. Particular thanks to Valtteri and Lilian for helpful comments on early drafts of the material. (In addition, my gratitude to Bruno Molder for looking at part of a very early draft.) I also appreciate advice from Tim Bayne about publishing the book.

Near the completion of the final draft of this book, I taught its content in a graduate seminar at the Centre for Time, University of Sydney, while a visiting fellow there. The small group of PhD students who attended that seminar helped me immeasurably, engaging with the material seriously and offering useful and clear criticisms and suggestions about the material at that stage; as such, my heartfelt thanks to Andrew Latham, Jane Loo, and Naoyuki Kajimoto. It wouldn't have been possible to present this material to them if I couldn't have visited the Centre. As such, I am also grateful to Kristie Miller, the director of the Centre, for the invitation and arranging the visit.

I would also like to thank Grace Weir, who initially suggested that I would better express my overall position on the relationship between time and experience in a single book (rather than spread it across several papers).

I am certain that I could not have persisted in this research without the encouragement and support of family and friends. I am greatly indebted to my parents, Nuala Power and Sean Power, for their unwavering trust in me. I am similarly grateful to my siblings for their support (especially Gerard for inspiring my interest in this area) and Paul Looby for suggesting that I keep at it.

Finally, my thanks to the editorial team at Routledge for their work which has led to a seamless and easy editorial and publication process.

1 Introduction

When I was very young, I shared a room with my much older brother. I was afraid of the dark. I would fall asleep with the side lamp on. Sometimes, I would wake up later, in the dark. My brother Gerard would have gone to bed in the meantime and switched out the light. I felt quite safe for most of the time we shared the room.

But one night, I woke up and the light was still on. My brother was reading. I sat back abruptly and struck my head (or something like that). A moment later, I was looking up at the ceiling. It appeared as if purple spots swarmed there.

"Gerard," I said, in a low voice.

"What?" he said, bored.

"There's something moving on the ceiling."

He paused for a moment. Then he closed his book and leaned over.

"They're gremlins, Sean," he whispered. "Don't move or they'll get you."

He switched off the light.

1.1. Trustworthy Experience

Can we trust our experiences? Can we hold them as a source of knowledge about the world? Can experiences (or *appearances* or instances of *phenomenology*) ground our beliefs so that beliefs based on them are instances of knowledge?

Based on some events in my childhood, I want to deny that experiences can always be *trustworthy*.[1] I do not want it to be that there were strange creatures on the ceiling, even if it appears that way to me. At the very least, I want to deny some descriptions and explanations of that experience, such as those given by my brother.

Furthermore, in a broadest set of possible worlds, there are reasons for holding that we cannot trust experience. It is possible that a subject is having deceitful or erroneous experiences. For example, the subject could be dreaming or fooled by an evil genius, a brain in a vat, or a battery in the Matrix. In all these cases, how things appear—particular objects and events in the actual world around the subject—is not how things are. The appearance of actual, particular objects and events occurs in the absence of those actual, particular objects and events.

In response to this problem from erroneous experiences, some epistemologists advance *externalist* theories of knowledge. If the processes that give rise to a belief are reliable, then even if the believer has no awareness of what makes that belief reliable, that belief can be knowledge. So even if we do not know if the processes giving rise to our experiences are reliable or not, so long as we are in fact having reliable experiences, then any belief based on them is knowledge (for discussion, e.g., Pritchard 2016).

Similarly, in perceptual theory, perceptual theorists, *disjunctivists*, defend a view that there are 'bad' and 'good' perceptions. The good perceptions are veridical, the bad not. They otherwise have nothing in common, despite their appearing to have something in common—e.g., hallucinations are bad perceptions (or, even, not really perceptions at all)—yet appear to the subject as the same as veridical perceptions (for discussion, see Logue 2015).

However, these approaches raise a problem for the subject of the perception. They ignore reasons or justificatory factors which are internal—i.e., capable of being accessed by reflection or introspection. Under an externalist or disjunctivist analysis of experience, the subject cannot tell apart cases of knowledge and false belief or cases of good perception and bad perception.

This problem is particularly troubling in cases involving experience. An experience of x is an appearance of x. The appearance of something seems to be something that is accessible to us. In the epistemological sense, the appearance of something is *internal*; it is accessible to the subject; in their reasoning about the world, the subject can employ appearances.

Experience may be reliable in these cases and so a justified source of belief and knowledge. However, experience is reliable (and distinguishable from unreliable experiences) because of something inaccessible through subjects' experiences. It is not because of how things appear to us through our experiences. Instead, it is something else that does this work. As the subjects of perception, we may have a 'good' experience of the world but we cannot tell it apart from the 'bad' experience.

If we identify such an experience's trustworthiness with its reliability, this leaves an important problem untouched. For any particular experience, the question 'can I trust this experience?' goes unanswered. Only others can tell me if I can trust it—and even then, only if their knowledge of the world is itself trustworthy—i.e., if their own experience is a source of justification which is accessible to them (which it may not be).

These thoughts may promote the view that in working out what is true of the world, we should do so without trusting experience. We might say we should not be empiricists (or at least not naïve empiricists). We should be rationalists or, at least, develop a form of empiricism that does not have experience as an epistemological ground (the particular way of distinguishing these positions follows Musgrave 1993). Whatever such alternatives are like, in these models, knowledge of the world (including its grounds and method of acquisition) is independent of our experience. We might call experience *epistemologically epiphenomenal.*

Yet can we have knowledge of the world without some of it depending on experience? We may well have knowledge of what is necessary and even of what is possible. However, even with such knowledge, a problem remains. With experience, we may know what is true of any world, or what is necessarily true. However, without experience, it is not clear how we can know what is true of contingencies in the actual world.

1.2. Knowledge of the Actual World

One worry that motivates this book is that, without experience, we cannot know about an important class of actual things (including events and objects). These are *contingent* things: things in the actual world that are not necessary things. They need not be true in every possible world (e.g., there is a possible world in which stars and atoms do not exist). It seems that to know which of a range of possible worlds we are in, to pick out the actual things over the possible and necessary things, we need to rely in some way on experience. If we cannot rely on experience, then I am not sure how we can know about what is actual.

Knowledge without experience can be coherent; it can be general or particular; it can be knowledge of many possible worlds. I can know true things about possible creatures such as gryphons and chimeras, such as the differences between them, without ever having experienced a gryphon or chimera. However, although such knowledge is possible, it can still fail to apply to the actual world. Whatever it is I know about gryphons or chimeras, without gaining this knowledge through experience, I cannot know if any such creatures actually exist.

I think that this role for experience is one reason to prefer empiricism, even a naïve form of empiricism, to rationalism. Something like this worry is found in the thinking at least one early empiricist. Discussing the medieval theory of demonology (a theory of witches, magic, and demons), Montaigne was dismissive of it. However, it is not obvious that he dismissed it because it was irrational. According to the historian Clarke, demonology's claims are logically possible. The beliefs one holds in accepting demonology are coherent with each other. However, Montaigne objected to it because its rationality is insufficient justification for holding that the entities particular to demonology belong to the actual world. The problem with a theory based on reason alone is that

> what reason does "is ask how such a thing happens; what it should do is question whether it happens at all." "Reason invents an 'other'

> world that is rationally coherent, complete with reasons and causes, but without foundation or substance."
>
> [. . .] The result is that "we know the foundations of and causes of a thousand things that never were."
>
> (Clark 2007, 289; also, Montaigne n.d.)

Despite being a rational theory, a theory may not correspond to the actual world. More precisely, it may correspond to some things in the actual world, the things that are necessary—e.g., the identity of the smallest prime number—that is, what must be true in any possible world. Yet other unnecessary but possible aspects of the theory may not correspond to things that are *contingent*—i.e., what is *particular* to the actual world. If that is so, then, even given a rational theory, the following question remains. How do we know about contingent things in the actual world?

A naïve answer is that we either know about the contingent things in the actual world through their being perceptually *apparent* to us or by their being inferred from what is perceptually apparent to us.[2] That is, we know about the actual world based on perceptual *experience* (or, also, as I call it sometimes, perceptual *appearances* or the *phenomenology* of perception).

Perceptual experience is not merely perception. They can be independent of each other. Perceptual experience is what a phenomenologically rich perception is like. It is also what a convincing hallucination or illusion is like. It is what disjunctivists deny is shared by perception, on the one hand, and hallucinations or illusions, on the other. Perception has success conditions; we only perceive if we successfully do so. One can have unconscious perception without experience (e.g., Humphrey 2008)—e.g., you perceive when you flinch away from something, but you do not experience it.

The suggestion here is that perceptual experience matters by giving a particular access to the contingencies of the actual world. As such, to know these things, I have reason to want perceptual experience to be trustworthy.

Even if they are not explicit about it, philosophers of perception accept that perception and perceptual experience play this justificatory role. For example, in his introduction to an edited anthology dedicated to the philosophy of perception, Matthen refers to experience through perception (which I assume is perceptual experience):

> Perception is the ultimate source of knowledge about contingent facts. We know about our surroundings because we are able to experience them through perception; we know about scientific phenomena because they are observed. Epistemology is therefore very much concerned with the evidential value of perception. Analytic epistemology is concerned with the rational grounding that perception

> lends belief; empiricist philosophy of science erects the entire edifice of scientific knowledge on the back of perceptual observation.
>
> (Matthen 2015, 1)

Siegel and Silins write,

> Seeing a jar of mustard in the refrigerator can give you reason to believe that the fridge contains mustard. [. . .] When you see a jar of mustard, you have a perceptual experience [. . .] [W]hen experiences provide reason for beliefs, they justify them. [. . .] In this entry, we begin from the assumption that [perceptual] experiences [. . .] can justify external world beliefs about the things you see.
>
> (Siegel & Silins 2015, 781)

That is, for claims about the merely actual world and about contingent facts, reason alone is not sufficient to test or justify such claims. However, perceptual experience can do that. It can play that role of testing or justifying such claims. If that is correct, we can take perceptual experience as a way of testing one's theories about the actual world.

This does not mean that experience must be of everything contingent about the actual world. However, if taken seriously, experience picks out some of the contingent things, and we can say we know about these things. As for contingent things about which we have beliefs wholly independent of our experience, I am not sure what we can say about them. For example, say I believe that there are gryphons living in my neighbourhood, yet this belief has no grounding in any of my perceptual experiences. Gryphons are creatures resembling a lion with the head of bird; they seem to be something that I could perceptually experience. In the absence of such experience, then, even if it is true that there are gryphons, I do not know what else could make my belief justified.

This is a very simple picture of how one might use perceptual experience to test theories about the actual world. It involves a naïve theory of perception—a theory focused only on the actual world. It may not work for all possible kinds of experience (e.g., those possible experiences had by demon-fooled spirits or brains in vats), but at least we can trust it for actual kinds of experience.

Unfortunately, there are significant challenges to this naïvety about perceptual experience.

1.3. Erroneous Perceptual Experiences

The most important challenge to naïvety about perceptual experience is that there appears to be overwhelming evidence that subjects of perceptual experience are subjects of actual cases of what I shall call *erroneous perceptual experiences*. These erroneous perceptual experiences are not only *possible* cases of erroneous experience, as they might be in classic epistemological thought experiments (such as being a brain in a vat or a mind misled by an evil genius). These are cases of *actual* erroneous perceptual experience, with actual evidence in support of them.

There seems to be overwhelming evidence for actual erroneous experiences. Many of these experiences are common and easy to generate. There are many websites and picture books with perceptual illusions. (It is even easy for many people to draw some of them, given a good straight edge.)

If it is so easy to acquire such evidence of erroneous perceptual experience, then there is evidence that a perceptual experience of something is easy to generate in the absence of the actual thing. This provides good reasons to doubt that how things appear is trustworthy evidence of how things are. Furthermore, even if such erroneous perceptual experiences were uncommon (as hallucinations tend to be), a single instance of such erroneous experience is enough to show that appearances can separate from reality.

As such, there is reason to doubt that appearances are trustworthy, exceptionless grounds for claims about the actual world. For, if anything can seem to happen without actually happening, what reason do we have to treat that something seems to happen is a reason to believe that it is happening?

One reason to treat appearances as corresponding to reality is that one *wants* how things seem in perception to justify beliefs about how things are. Again, one wants to use experience to pick out what is true of the actual world.

However, desire is not enough to make it a perceptual experience trustworthy. What we need is a reason to trust experience, not just a desire to trust experience.

Although this desire is not a reason to trust experience, it can be a reason to prefer one theory to another. One can prefer the theory that enables experience to be the most trustworthy.

Furthermore, that one can trust experience may depend on what one holds to be true of the world. In turn, what one holds to be true of the world may depend on assuming a particular metaphysical theory or ontology—a theory or ontology that one need not (nor should not) assume.

I explore both of these possibilities in this book. I argue that, in light of plausible theories about time, some alleged instances of erroneous perceptual experience are not erroneous at all. Furthermore, even if they are erroneous, they may not be erroneous in the way that we may tend to understand them to be erroneous. Instead, we are mistakenly interpreting them as erroneous because one holds a mistaken metaphysical position or ontology.

It is important to note that, even if a subject possesses some metaphysical position or ontology, they may not know that they hold this position. They may hold metaphysical assumptions unexamined by their possessors. This is particularly the case if it is due to intuitive yet philosophically problematic positions—positions held by default by most people but rejected by those who theorise about them.

The false yet intuitive positions that I think drive interpretations of many (although not all) experiences as being erroneous are positions in the philosophy of time. The central argument of this book is that, if we put aside intuitive concepts of time and assume more physically compatible concepts of time, we can frequently reclassify experiences as veridical that we intuitively (at least, allegedly) interpret as erroneous.

1.4. Ontological Commitment

The metaphysical discussion about time of most interest here is the debate concerning (a) the reality of time and temporal properties and (b) the reality of what is in time or has temporal properties. Many philosophers of time hold that the positions in these debates have the following relationships to, on the one hand, intuition (and/or common sense) and, on the other hand, modern physical theory:

(a) *Presentism* and *tense theory*. These theories are intuitive.
(b) *Eternalism* and *tenseless theory*. These theories are counter-intuitive.
(c) Modern physical theory is more compatible with eternalism and tense theory than it is with presentism and tenseless theory.

How these positions feed into interpretations of perceptual experience as erroneous turns on their consequences for what one holds to be real—i.e., on the ontological commitments these positions involve. These commitments about reality might include what is possible but, importantly, also include what is *actual*. What I argue in this book is that different positions on time, applied to the actual world, have different consequences for one's ontological commitments about the actual world. As such, through their impact on one's ontological commitments, different positions on time have different consequences for interpreting any actual perceptual experience as being erroneous.

Generally, if one has a perceptual experience that seems to be of something, then whether or not it is veridical or erroneous depends on whether or not there is, in actuality, that thing. Furthermore, it is the case that, if one has a perceptual experience that seems to be of something, then whether or not we *believe* it is veridical or erroneous depends on whether or not we believe there is, in actuality, that thing.

As an example about something independent of time, I draw on another childhood event. When I was very young, many people in my county claimed to see religious statues moved by divine intent. Some even saw specific religious figures, such as my neighbour claiming to see Saint Anne (my neighbour claimed they knew it was Saint Anne because she resembled a renaissance portrait of Saint Anne).

To these people, these were obvious miracles. Their ontology included religious miracles; as such, they did not think that their visual experience was mistaken. However, for me, seeming to see purple spots move on the ceiling, my ontology does not include these purple spots and certainly not that they are gremlins. As such, I believe my brother was deceiving me (which may not surprise the reader).

Similarly, in this book, my ontology includes several things and excludes several things that influence my beliefs and arguments. I argue that thinking in terms of one position on time when an alternative position is true can falsify one's interpretation of perceptual experience as erroneous. I discuss several philosophical positions about time, but I do not fully consider experience from all these positions. I use many to illustrate how switching between positions on time can change how one models experience in the world, and as a result illusion and hallucination. I also bring in positions I do not commit to because they are allegedly intuitive and may influence the reader who is not a philosopher of time.

In later chapters, I discuss these positions in more detail. However, since it affects the entire presentation of this work, it is important to explicate briefly my own ontological commitments.

With respect to the philosophy of time, I assume the following:

- The allegedly intuitive positions of *presentism* and *tense theory* are false. Their opposing positions *eternalism* and *tenseless theory* (as motivated by the *special* and *general theories of relativity*) are true.
- I am justified in my assumptions because well-confirmed actual physical evidence supports them.
- The actual world is one in which the special and general theories of relativity are true.

With respect to perception and experience, I assume that

- Experiences happen in time and in space.
- Physicalism is true of the structure of experience. In particular, it is true of the spatial and temporal structure.

As such, my target audience accepts contemporary physical theory, is interested in the consequences of the metaphysics of time, and is physicalist about the spatial and temporal structure of perceptual experience.

I think that investigating such ontological commitments is important. It is my general position that it is a desideratum of any theory of perceptual experience to avoid an error theory or a surfeit of possible (and actual) erroneous experiences. This is an epistemological and empirical desideratum: The theory that more closely meets it enables perceptual experience to play a stronger role in showing us the actual world.

I do not know if this is a metaphysical desideratum. One may argue that metaphysics concerns precisely that which we cannot decide or investigate through experience or empirical research. As such, metaphysically, the role of experience or the empirical may be weak or irrelevant. Furthermore, from a wholly metaphysical standpoint, the aforementioned desideratum does not mean that the theory contains a more accurate or true description of the actual world. The more accurate or true theory may be one in which perceptual experience is commonly erroneous, unreliable, and untrustworthy.

However, in the absence of an alternative candidate for this role (of showing us the actual world), I take it that such a theory has significant problems about knowing what is true of the actual world. This may not concern metaphysicians. However, I take it to be a concern of many other theorists and so a good motivation for looking at the consequences of ontological commitment from the philosophy of time (just, perhaps, not the philosophy of time for metaphysics).

Similarly, for time, ontology, and experience,

1. The concept of time to which one commits (partially) determines one's commitments to an ontology.
2. One's commitments to an ontology determine
    - One's commitments to the truth or falsity of statements about the world.
    - One's commitments to whether or not a certain experience is erroneous.
3. If physicalism is true about the spatial and temporal structure of experience, then the physical concept of time captures the spatial and temporal structure of experience.
4. It is common to say that the physical concept of time is counter-intuitive. It does not fall under common intuitions about time.
5. If a) one's assumptions about the relationship between experience and time are based on common intuitions about time and b) the physical concept of time is correct, then one's assumptions about the relationship between experience and time may conflict with the correct account of the spatial and temporal structure of experience.
6. If one's assumptions about the relationship between experience and time may conflict with the correct account of the spatial and temporal structure of experience, one may be mistaken about what constitutes a veridical or erroneous experience.

7. If one is mistaken about what constitutes a veridical or erroneous experience, then one is mistaken about what cases of experience are erroneous experiences, or how particular erroneous experiences are mistaken.
8. If one is mistaken about what cases of experience are erroneous experiences, or how particular erroneous experiences are mistaken, one can be mistaken about whether or not experiences fall under the categories of illusion or hallucination.

That is, one's assumptions about perception, especially erroneous experience, can be dependent on what one thinks about time. If one's thinking is mistaken, one's assumptions can diverge from what is actually true of perception, illusion, and experience.

Ontological commitment is important for perceptual experience and its epistemological role in deciding erroneous and veridical experiences from one another. It is also important in deciding the kind of error that a particular experience is.

For example, having an experience as of a divinely moving statue (as many shrine visitors claim to see), I may deny miracles but accept instances of incomprehensible technology. Thus I accept that my experience of the statue moving is veridical, but any experience of divine presence or proclamation associated with the statues is not veridical. I may accept purple spots on the ceiling, but not that they are monsters.

However, to show that we can make these selections in terms of kinds of error, we must first show that these kinds of error are distinct. Thus part of this book concerns itself with distinguishing kinds of error. I argue here that we have three possible kinds of error: illusion, hallucination, and anosognosia.

The concept of anosognosia as I develop it here is new in the philosophy of perception. I adapt it from psychology, generalising from part of a pathological perceptual condition to general perception.

Briefly, anosognosia is an error related to perceptual experience. If a subject is anosognosic of X (whatever X might be), then there is an X, yet X is not apparent to the subject. Furthermore, there is no further sense to the subject through experience (such as that provided in amodal completion) that there is any such X. For the subject, the result of this anosognosia is that something that is the case appears to be not the case—e.g., there appears to be no X.

Furthermore, we can use differences in kinds of erroneous experience to sort between theories. In Chapter 5, I argue that, for picking out facts about the actual world through one's experience, commitments to some kinds of erroneous experience are worse than commitment to other kinds of erroneous experience. Some kinds of experience involve a greater divergence between what is apparent and what is actual, leading to a greater amount of error and untrustworthiness of experience. In this respect, I argue that a) a theory that commits one to hallucinations is worse than a theory that commits one to illusions or to anosognosia and b) a theory that commits one to illusions is usually worse than a theory that commits one to anosognosia (the exceptional cases being where both kinds of error are interchangeable). This is because anosognosia puts one in an epistemologically more positive and simpler position than illusion or hallucination, and the same applies to illusion in comparison to hallucination.

Finally, some theories of time commit one to the worse kinds of erroneous experiences, but others do not. As such, we can use the kind of erroneous experience to which different theories of time commit us to sort between those theories.

Thus if one commits to the wrong theory of time, then one mistakenly interprets some experiences as being erroneous or, even if it is correct to interpret them as erroneous, as being the wrong kind of erroneous experience. Furthermore, one should prefer theories of time that commit one to veridical experiences or to certain kinds of erroneous experiences to others.[3]

1.5. Time and the Perceptual Experience of Space

To illustrate the arguments in this book, I consider a fundamental form of perceptual experience, the perceptual experience of spatially located things. The experience of space is fundamental because almost all, if not all, of what we experience seems to have a spatial location. This includes all modes of perceptual experience—e.g., anything experienced visually, auditorily, tactilely, or proprioceptively appears to be spatially located, either within or outside the subject's body. Furthermore, much of the identity and properties of things depend on those things' being located in space. For example, what appears to be a tree is not a tree if it is only two-dimensional, or has no spatial location at all. What might appear to be a distant explosion is not a distant explosion if it is merely a tiny sound located within the confines of the ear (e.g., in the vibration of filaments in headphones).

However, the perceptual experience of space also seems to be particularly susceptible to error. There are many alleged cases of erroneous perceptual experiences of spatially located things. There are geometric-optical illusions—e.g., the Ebbinghaus illusion—and mirages. There are hallucinations of things located at a distance from the experiencing subject—e.g., seeing your long-deceased grandmother in a rocking chair. There are what we might call *projections* of internal *sensa*—e.g., after-images or the purple spots I saw 'on the ceiling'.

I argue here that each of these may be cases of erroneous perceptual experiences. However, to hold justifiably that they are erroneous and, if they are erroneous, how they are erroneous, depends on what is true of time. The veridicality of some perceptual experiences of space depend in part on the temporal structure of the perceptual experience.

What about the perceptual experience of time? Is this not as fundamental an experience and as important as (or even more important than) the perceptual experience of space?

I agree the experience of time is important. However, there are two reasons I put that aside here.

First, the point of this volume is to demonstrate how time can alter thinking about perceptual experiences in general. It is not just about the experience of time. There are arguments about time and experienced time that are particular to it, and inapplicable to general experience. Time is both content and constituent in those cases. The arguments around the experience of space are better illustrations of how we can extend thinking about time to experience generally.

Second, we can divide the experience of time in two: perceptual experience of time and non-perceptual experience of time. In this book, I am only interested in perceptual experience. However, many interesting questions about temporal experience concern non-perceptual experience—e.g., anticipation and memory.

Third, a discussion on temporal experience and the philosophy of time is enough to fill at least one volume on its own. As I want to show how thinking about time works with experience in general, to include it would either require abandoning that project (focusing only on time) or else making a much larger volume.

This does not mean that I ignore the experience of time. Where it is relevant, I discuss it. One chapter is dedicated to it. However, I do not consider that chapter to be in any way exhaustive in its discussion about the subject. Instead, the overall theme of this book about perceptual experience and time determines the specific content of the chapter. It focuses on how thinking about the philosophy of time can inform thinking about the trustworthiness of temporal experience.

1.6. Scepticism about This Project

Two thoughts may promote scepticism about this project. The first is that the project is pointless: The centrality of perceptual experience for beliefs about the actual world is problematic whatever our position on time. The second is, given that there is a dispute about time on the metaphysical level, with respect to experience, one should abstain from taking a position on it. As such, one should ignore differences in the different positions about time in thinking about experience.

1.6.1. Perceptual Theory and Specific Perceptual Content

I do not think analysis of perceptual experience in light of different positions about time saves all appearances. Some are going to be in error, irrespective of what is true of perceptual experience in time. Even if we can understand some of these erroneous experiences as cases of anosognosia (which I argue is the preferable result), I think there are a remainder that must be illusions (in particular, erroneous experiences of depth) and hallucinations.

This raises the first worry. This project is irrelevant to questions about the veridicality of perceptual experience and its role in our knowledge of the actual world. It is irrelevant because it addresses particular cases of experience and error. Yet for perceptual experience to play its particular epistemological role, all instances of perceptual experiences must be veridical or none of them can be trusted to be veridical.

As such, this thinking might go, unless my analysis shows that some positions on time make all perceptual experiences veridical (what I shall call a *universal veridicality*), then there is not much importance to difference in accounts of perceptual experience under one position of time and another. We are still in as much of a position of being unable to trust experience to tell us about the actual world as we were when we ignored the philosophy of time.

I accept that there are actual cases of erroneous perceptual experience. I only argue that, given the most compelling position on time (tenseless theory), some alleged cases of erroneous experience need not be erroneous (or are not particular kinds of erroneous experience). For experience to play its desired epistemological role, if only universal veridicality works, then my approach here does not work.

However, I do not see it as obvious that universal veridicality is necessary for questions about the trustworthiness of experience. Universal veridicality is only necessary for the trustworthiness of all *instances* of experience. What I am interested in is the trustworthiness of *some* instances of experience. My position is that some cases of experience can be trustworthy and always veridical.

1.6.1.1. *Disjunctivism and Trustworthiness*

This position may seem like a standard disjunctivism, with its 'good' (here, trustworthy cases) cases of experience and 'bad' (untrustworthy) cases of experience. However, it is not exactly like standard disjunctivism. I divide the trustworthy from the others in terms of what is experienced—i.e., what may be called the content.[4] The difference is roughly this: We (actual subjects) can have experiences of some things only if they are actual (are in the *actual* world); these are trustworthy experiences. We can have experiences of other things even if they are not actual (i.e., they are not in the actual world); these are untrustworthy experiences.

Unlike standard disjunctivism, my line of separation is between experiential contents for which there is genuine evidence of erroneous cases and those for which there is not. The main difference between my theory and disjunctivism is that mine is falsifiable by actual empirical evidence.

It is possible that there is genuine evidence of erroneous experiences for all contents. If that is right, then, as I present the term here, there are no trustworthy cases. However, there may still be 'good' and 'bad' cases, as conceived by standard disjunctivism.

As such, my justification for trusting particular experiences is dependent on empirical evidence. This may raise familiar epistemological problems about experience—e.g., of circularity (e.g., see Pritchard, *op. cit.*)—but, for those particular experiences, it does not raise the problem of error—unreliability, untrustworthiness, illusion, or hallucination.

Along with the perceptual experience of time, the kind of content for perceptual experience I am interested in is that of space. In later chapters, I argue that experiences of the externality of things can be veridical and even trustworthy given some positions on time, and cannot be either veridical or trustworthy given others. However, even where the experiences of externality are trustworthy, the depth or distance from experiencing subjects of what is experienced is frequently false, irrespective of the truth of different positions on time.

Finally, dividing perceptual experiences this way is somewhat motivated by the practice of scientific investigations into perception. Science does not tend to treat perceptual experiences overall as being undermined by any specific perceptual error. Relevant perceptual errors undermine an experience, not just any perceptual errors.

This is clear by considering what follows from the truth of the universal position. Say that the universal position is correct; if I wonder if my perceptual experience of colour is trustworthy, then sufficient evidence of untrustworthiness is that I sometimes experience things as having the wrong *weight*. Given the universal position, my mistaken experience of weight means that no experience is trustworthy.

That this is sufficient is not obvious to me. It seems to me that, in practice, theorists do not treat it as obvious either. For example, in practice, researchers give examples of illusory colour to show that we cannot trust our experience of colour. Illusions of weight do not count towards that. We might say that, in practice, illusory colours are *relevant* and illusory weights are *irrelevant* to the trustworthiness of experiences of colour.

What makes a certain illusion relevant to the trustworthiness of an experience? The answer, again in practice, is *content*. If a particular experience has the same content as an illusion, then that illusion is relevant to that experience.

Indirectly, illusions of other content can be relevant to an experience. However, we should not assume this relevance; there must be a reason for it. For example, there might be some connection between one property (e.g., a particular colour) and another property (e.g., a particular weight) such that the absence of one is an absence of the other (e.g., if there is no instance of that particular weight, then there is no instance of that particular colour). As such, if there is an appearance of the former property (that particular colour) in the absence of the latter property (that particular weight), we have a reason to hold that there is an illusion of the first property (an illusion of that particular colour).

In the chapter on illusion, there is a general discussion about such connections. For now, it is sufficient to note that I deny that, in the context of perceptual theory and thinking about the actual world, evidence for some erroneous experiences is insufficient to deny the trustworthiness of other experiences. As an example of my thinking, in later chapters, I consider the perceptual experience of space. We may wonder if the perceptual experience of space is trustworthy, and so if we may take it as evidence of actual space. To show that it is not, we show that there is evidence of erroneous perceptual experiences of space. We do not show it with other erroneous perceptual experiences, such as of weight or colour. (Unless, that is, one cannot have instances of the latter errors without instances of erroneous experiences of space.)

In response, one might hold that this is a mistaken approach to erroneous experience. If there is any case of erroneous perceptual experience, then all cases of perceptual experience are untrustworthy. If this means there is some possible world where all cases of perceptual experience are untrustworthy, then I agree. However, as originally discussed, when it comes to explaining actual experience, I do not think including experiences from merely possible worlds is necessary. As such, I do not think including erroneous experiences from merely possible worlds is necessary. If my thinking here is right, and one thinks it is necessary, then one must argue for why it is necessary.[5]

1.6.2. *Ontological Abstinence*

The second worry concerns what I call *ontological abstinence*. Ontological abstinence is the position of neither committing to nor rejecting an ontology. One *abstains* from committing to either being for or against the ontology.

For example, there is an ontological dispute between *modal realists* (or *possibilists*) and *ersatzists* about possible worlds. Modal realists hold that merely possible worlds are as real as the actual world; ersatzists hold that merely possible worlds are not as real as the actual world (e.g., Melia 2003).

If one ontologically abstains from this debate, one abstains from committing to either position. One neither takes the position that possible worlds are real nor takes the position that possible worlds are not real.

Ontological abstinence is justified in some cases. It is justified when two competing positions are equally matched (in some sense or other) in their accounts of a subject. For such a debate, one cannot appeal to authority about the disputed ontology in holding a position, and one needs to abstain from holding a position.

In theory and practice, one might then ontologically abstain from a position. For example, in formulating her theory, it seems right that a biologist ontologically abstains on the substance debate about space (between substantivalists and relationalists). She should not assume that the substantivalists are correct, and she should not assume that the relationalists are correct.

However, ontological abstinence has an important consequence, especially for ontology in the philosophy of time. If one persists in ontologically abstaining from a debate, one must formulate theories that are as compatible with one side of the debate as they are with the other. Being as compatible with one side as the other requires one to restrict one's ontology to what both sides have in common.

For example, substantivalists and relationalists about space agree that there are things that are not merely reducible to aspects of time or space. As such, the biologist can assume in the formulation of her theories that there are things that are not merely reducible to aspects of time or space. However, the biologist cannot assume that there is space independent of those things. That is not consistent with relationalism.[6]

Ontological abstinence is relevant to the philosophy of time because one might think that one should abstain from taking a side on disputes in time. For example, in the presentism/eternalism dispute, one should not take either the presentist or eternalist side. So, one restricts oneself to the common view of both positions—e.g., that there are real things in a single moment in time, the present (be it relative or absolute). Beyond that, eternalists and presentists part ways, with eternalists holding that there are real things at other times as well.

Thus, similar to the biologist theorising without a commitment to the substantivalism dispute, when one abstains from the dispute, all theories one provides need to be restricted to real things being at only that single moment in time. Holding that there are real things at any other time is taking a side and not abstaining.

For example, if an abstainer theorises about perceptual experience, then his or her theories about the constitution of perceptual experience and the reality of perceptual experience's contents must assume only that one moment, the present, is real.

However appealing this may be, I think one should not abstain from the debate in the philosophy of time. Even if one does, I do not. I assume one view and, in my theorising about the structure of things, including perceptual experience, I run as far as I can with it. My aim in this book is to show the results of doing this.[7]

1.7. Chapter Outline

The book begins by outlining the philosophy of time. This is an important yet not commonly understood area of philosophy, assumptions about which can determine thinking about experience and error. I then move on to discuss experience and error. Finally, I bring these topics together and discuss the results.

Chapter 2 introduces the philosophy of time, describing the various relevant positions and the relevant arguments for them. As stated earlier, I assume the positions of eternalism and tenseless theory are true, those of presentism and tense theory false. However, it is important to make all these positions explicit here, as their differences may not be obvious to those not engaged in the philosophy of time. Furthermore, how each conceives of reality feeds directly into thinking about experience and error.

Chapter 3 outlines my view of the structure of perceptual experience. I distinguish perceptual experience from experience in general and outline the structure of perceptual experience. To do this, the chapter draws on concepts of constitution from extended mind debates. In these debates, the relevant concern is with the constitution of mental entities over space. I extended this here to the constitution of entities over space and time, and then narrowed it to the constitution of perceptual experience over space and time.

Chapter 4 outlines the different kinds of erroneous experience: illusion, hallucination, and the new kind introduced here, anosognosia. I argue that the distinction between illusion and hallucination is not obvious. It involves some metaphysical or ontological commitment—e.g., to distinguish hallucination from a complex illusion of properties, one must assume objects can be distinguished from bundles of properties; this is a metaphysical view. I also define anosognosia here.

Chapter 5 is an argument that some erroneous experiences are preferable to others. It is a good-making feature of a theory that it is committed to the preferable error over the less preferable error. I argue that a theory committed to illusion is preferable to a theory committed to hallucination, a theory committed to anosognosia is preferable to a theory committed to illusion, and this preferential order is transitive (i.e., a theory committed to anosognosia is preferable to a theory committed to hallucination).

In this chapter, I also argue that we can reconceive or reduce some kinds of alleged cases of illusion to cases of anosognosia. However, we cannot do this with hallucination.

In the chapters that follow, I turn to the perceptual experience of space. They concern the relationship between things in time and space, and then focus on the relationship between perceptual experience and spatial things.

Chapter 6 is an examination of the spatial relation between present and past things. I argue that different views of time deny such spatial relations; others accept and even promote them.

Chapter 7 looks at perceptual experiences of depth in light of different theories of time.

Chapter 8 looks at the kind of error in the perceptual experience of space. It looks at the conditions for a distorted perceptual experience of space given one's understanding of perceptual experience as extended in time.

Chapter 9 examines the experience of time itself—i.e., temporal experience. First, I consider the possible content of temporal experience and the possible errors related to such content. In keeping with the overall theme, I argue that the errors in such experience vary depending on what is true of time. Second, I look at models of temporal experience, in particular the retentional and extensional models. I argue that the plausibility of each of these models depends on what is true of time.

Finally, *Chapter 10* is more speculative. I consider how we might eliminate hallucination. In previous chapters, I argue commitment to this kind of error is worse than commitment to others. I consider how we might avoid interpreting some actual experiences as hallucinations. I offer a possible way of doing so given tenseless theory and, in light of this theory, the extension of perceptual experience in time.

I consider this book as the beginning of a project with at least one ideal outcome. This outcome is better support for the veracity of actual experience. I argue that we get it through thinking differently about experience in time.

Some might doubt the feasibility of this project. Even so, as a final point, here is why I think it is worth attempting.

First, as discussed earlier, I would prefer to hold a position about the world from which experience can be trusted and can be taken as it appears. Failing that, at least a world where experience is more trustworthy than any alternatives. This project is an attempt to show how different theories of time change the trustworthiness of experience.

Second, an ontological and constitution-directed discussion lays out how experience is in error, with the possibility of formulating some general rules and predictions from those rules. This work explores how experience can be in error, rather than simply assuming that it can be. It is not about observing error and then leaving the explanations of the error to later work. This is about describing the error, getting as clear as possible about the ontological commitments in that description, and seeing if there are other ways to describe the experience without ascribing error to it.

Finally, I have no philosophical qualms about denying that there are actual cases of kinds of erroneous experience. I do not think that we must *save* erroneous experience. Unlike veridical perceptual experience, there are no epistemological problems that I can see with, say, eliminating illusions and hallucinations. If it turns out there is a theory of time (or any other kind of theory) that preserves all experiences, making all cases of appearance correspondent with reality, then I assume that is a good feature of that theory, not a bad feature.

I do not think that such a theory is possible. There is good evidence for actual cases of erroneous experience. For example, as I discuss later, no matter which theory of time is accepted, I do not think that we can eliminate illusions of depth. However, I do think many alleged cases of illusion and hallucination are not so convincing under the theories of time most likely to be true. Many are still erroneous, but they are anosognosic. I think this is better than being committed to illusions or hallucinations.

To show this, I begin with the philosophy of time.

Notes

1. Throughout this work, I refer to the *trustworthiness* rather than *reliability* of perceptual experience. I think that something reliable is something that can go wrong. It is reliable because it is only *usually* right. I am interested in something that cannot go wrong. That something is what I refer to as *trustworthy*.
   I distinguish *trustworthiness* from *reliability* as follows:

   - An experience of E is *reliable* if, given the experience of E, the probability of E is greater than chance. Alternatively, a reliable experience of E increases the probability of E.
   - An experience of E is *trustworthy* if, given the experience of E, then E. Alternatively, a trustworthy experience of E gives a probability of E that is 1.

   As is perhaps obvious from this distinction, a thesis that an experience of E is trustworthy is much stronger than a thesis that the experience of E is reliable.
2. Such inferences would be through theories that are virtuous in that they are economic, simple, and have sufficient explanatory force. These virtues may not be directly connected to experience, but theorists often appeal to them in accounts of knowledge about unobservable things (my thanks to an anonymous reviewer for raising this point).
3. It may help to think of this discussion in terms of the Duhem-Quine thesis in the philosophy of science (e.g., Ladyman 2002). Different theories lead to different descriptions of experiences, with different 'auxiliary' statements affecting the empirical status of such experiences. Some auxiliary statements lead to the experience being erroneous (and so to not being an epistemological foundation for evidence); others do not. My point here is that, from an empirical point of view and regarding questions about the actual world, the preferred theory in such cases is the one with the least cases of erroneous experience.
4. As such, one might call this a form of *content disjunctivism*. However, the phrase 'content disjunctivism' already commonly refers to *disjunctivism* proper—i.e., the perceptual theory that divides indiscriminable perceptual experiences into good and bad, with the two kinds held to have nothing in common.
5. An anonymous referee notes that the 'traditional' Cartesian argument seems to be general precisely in that sense. One should not trust what has lied to one at least once in the past; therefore, one should not trust the senses. This applies as much to experience as the senses, especially perceptual experience
   At the epistemological level of Cartesian doubt, I agree with this point. Given all possible worlds, perceptual experience are untrustworthy. For example, it is possible that I am having the exact same experiences I have in this actual world, and yet I am merely a brain in a vat. In that case, arguably, all my experiences are non-veridical, illusory, and untrustworthy.

However, the context of this argument is not at the level of all possible worlds. It is limited to the actual world and the set of possible worlds compatible with actual empirical evidence. Furthermore, Cartesian doubt is a sceptical position, with its claims supporting scepticism. I take it that all the positions discussed here—both from perceptual theory and from the philosophy of time—are realist (or idealist), and not sceptical. When presenting my analysis here, my opposition is not sceptics about perception and time, but realists (or idealists) about them.

Another way of explaining my position is this: say that, whatever we can know, we are in fact *not* brains in vats (I happen to think this is the truth). Then why must a theory or analysis of actual perceptual experience be consistent with our *being* brains in vats? If such a theory or analysis need not be consistent with being brains in vats, then it need not be consistent with the consequences of being brains in vats, such as that our brain-in-vat experiences can be false and so cannot be trustworthy.

6. In addition, if she wishes also to abstain from a debate involving super-substantivalism about space, she cannot assume that there are things that are irreducible to space.
7. How are ontological abstinence and scepticism related? I take it the difference is an attitude about what we can know. Sceptics about a debate do not think we can justifiably hold either side of a debate—e.g., a sceptic about the presentist/eternalist debate does not think we can justifiably be either a presentist or eternalist. An ontological abstainer does not need to take this position about a debate. It is only that they think they do not need to take a side in the debate. (Of course, a sceptic would also think he or she does not need to take a side, as no one can take a side.)

2 The Philosophy of Time

> Things that happened in the past
> Only happened in your mind.
> (Biff Rose, 'Fill Your Heart')

If some philosophers of time are correct, the claim in Rose's song that there is no past is intuitive to most people. The intuition is there is no past *really*, or any time other than the present. Certainly, we talk and think about other times in the past (or the future). We draw diagrams of them, timelines, and quantify over them ('the evolution of the mammalian eye occurred millions of years ago'). However, earlier times than the time you are reading this have no actual reality or existence. They are not anywhere 'out there' in the world. Only the present is real. Only what happens *when* you are reading this is *really* happening.

This may be the intuition behind those particular lines. However, there are good reasons to doubt the underlying metaphysics of these intuitions. There are good reasons to hold that various things—which includes events, processes, objects—at earlier or past times to your reading this are as real as your reading this; things at future or later times than your reading this are as real as well; what is at earlier and later to your reading this are 'out there'.

More than that, it is argued here that, with respect to the structure of experience, there is good reason to hold that times other than a *strictly present* time are 'in here'—that is, are constituents (and not just contents) of experiences, even of your experience of reading this.

These good reasons come of the result of three significant temporal debates:

- On the reality of objects and events in time
- On the nature of real or genuine change—and as part of this, the reality of tense and the meaning of 'dynamic' and 'static'
- On time and contemporary physics

These debates about time have led philosophers to develop and endorse various metaphysical positions on time. For this book, the most relevant debates (and sides in these debates) are as follows:

(1) Presentism versus eternalism
(2) Tense theory versus tenseless theory
(3) Relative simultaneity versus absolute simultaneity

One way to understand the different positions in these debates is through their positions on the nature of the past, present, and future.

As I discuss in a later chapter, according to presentism and tense theory, these distinctions are absolute and fundamental to the nature of time. In particular, the present or *presentness* is unique or singular, as well as being absolute and fundamental. There are not multiple presents. Presentness is not relative to or derived from something else (e.g., Lowe 2003).

In contrast, according to tenseless theory, the past, present, and future are not absolute or fundamental to the nature of time. There is no unique or singular present. There can be multiple presents. Presentness can be relative or derived from something else (e.g., Power 2012).

Throughout this book, I demonstrate the importance of this difference through its effects on thinking about experience in time.

An immediate effect of this difference is in their different conceptions of *real change*.

2.1. Real Change

Tense theorists and tenseless theorists disagree about what is necessary for real change. In brief,

- For tense theorists, real change is temporal passage (what I sometimes call *tense change*).
- For tenseless theorists, real change is variation over time in the properties of things (what I sometimes call *tenseless change*).[1]

Temporal passage is a change in time itself. It is (metaphorically, at least) like the flow of water in a river. However, a river's change is a kind of motion—i.e., a change of something's position in space (e.g., a wave of water's location above a spot on the riverbed). With time, the change is not of positions in space. It is change in something else.

With temporal passage, what changes? What is the analogue of the flowing water in a river? In general, what changes is something's *position* in time.

Beyond that, there are various conceptions of temporal passage. Here are the two most important.

2.1.1. Changing Events and the Moving Spotlight

The first concept of passage is of the *events* changing.

I am sitting in a dentist's office, waiting to travel, or waiting to stop travelling. I am just passing time, doing nothing, my surroundings and myself seemingly unchanging. The event that brought me here (a stabbing toothache last night, a taxi to the airport) is receding into the past. The event that ends the wait (being called in by the dentist, getting on the plane, getting out of the taxi) approaches, with no obvious effort needed on my part. The event comes from the future, it will happen, whatever we do, whatever we do not do.

This first conception of temporal passage (or tensed change) is that events change positions in the *A-series* or *tenses*. The A-series (originally dubbed by the metaphysician McTaggart) is the series of positions running from the far past, through the recent past, through the present, through the near future, to the far future. Events gain and lose their positions in this series in a set order: Events are future, then they are present, and then they are past.

Philosophers commonly referred to these positions as *tenses*. As such, we can also refer to this A-series as a *tense* series divided into the past *tense*, present *tense*, and future *tense*. As such, in this conception, temporal passage is a *tense* change in the events.

One aspect of this conception of temporal passage is that the present itself does not change. The present is fixed in time, as indeed are the other A-series positions in the past and future (e.g., two weeks past, a thousand years in the future). The events change; they 'move' through time, moving through the otherwise fixed A-series, gaining and losing positions in it.

When distinguished from the one that follows, this concept of temporal passage is not the explicit viewpoint of any current A-theorists. However, it resembles McTaggart's view of what is necessary for change. According to McTaggart, for there to be real change, the *events* must change, and they must change their position in time. Responding to Russell's account of change, which does not involve temporal passage, McTaggart writes,

> Mr Russell looks for change, not in the events in the time series, but in the entity to which these events happen . . . [I]f my poker, for example, is hot on a particular Monday, and never before or since, the event of the poker being hot does not change. But the poker changes, because there is a time when this event is happening to it, and a time when it is not happening to it [. . .]
>
> But this makes no change in the qualities of the poker. It is always a quality of that poker that it is hot on that particular Monday . . . [and always a quality that it] is not hot at any other time . . . The fact that it is hot at one point in a series and cold at other points cannot give change, if neither of these facts change—and neither of them

> does. Nor does any other fact about the poker change unless its presentness, pastness, or futurity changes.
>
> (McTaggart 1927, 14)

The next conception of passage shifts the change to something other than the events; it moves it to the *present*. That is, the *present* changes. Like a pen across a page, or a spotlight across a row of houses (to use Broad's example), the present moves towards some events and away from other events.

Philosophers sometimes call this the 'moving spotlight' theory of time (e.g., Broad 1938; Skow 2012; Deasy 2015). The idea is that the events do not change in time but are fixed in it; the *present* changes. Then, by the present 'falling' on, 'reaching', or 'passing' the events, the events become present. By the present doing so, events that are earlier than those events become past (or more past, if already past) and events later than those events become present (or come closer to the present) from the future.

Whichever of these concepts of passage one holds, they include the same assertion by philosophers who defend temporal passage. Philosophers who defend it, often called A-theorists or tense theorists, argue that this *temporal passage* is necessary for time. If you deny the reality of such passage (as other philosophers of time, tenseless theorists, do), then, claims the tense theorist, you deny the reality of time.

This position about real passage, real tense, and real time has one of its main sources in the thinking of the philosopher McTaggart. In his paper, 'The Unreality of Time', McTaggart argued that time is unreal exactly because change in the A-series is impossible (McTaggart 1908). That is, McTaggart denied time is real because temporal passage could not be real.

One reason that one may deny that time (and passage) is real is that one denies that distinctions in the A-series are real. For example, one result of relativistic physics is that the past, present, and future are not distinct in reality. If this is right, then relativistic physics means that time and temporal passage cannot be real. This is because, if the distinction between past, present, and future is not a real distinction, then there is no real change between them—i.e., there is no real passage. If passage is not real, then, according to tense theorists, time is not either.

2.1.2. Tenseless Change

The next concept of change comes from theorists who do not accept that the denial of passage is a denial of time.

Those who do not hold temporal passage to be (or necessary for) real change are often called *B-theorists* and *tenseless theorists* (or, sometimes, *detensers*). Part of their defence of time without passage is that there can be real change (and time) without real temporal passage. This is because real change is not the same as a change in the A-series. It is change in properties of objects and events such as colour, shape, or spatial location.

For the B-theorist or tense theorist, variation between different times is sufficient for real change. For example, an apple is wholly green on Monday *before* it is wholly red on Tuesday. This only requires that something possesses a property at one time (it is wholly green on Monday) that it does not possess at another time (it is not wholly green (but wholly red) on Tuesday).

There is another important aspect to this account, an aspect that gives the name to B-theorists and tenseless theorists. This is the meaning in the description of this change of the phrases 'one time' and 'another time' or, more generally, the meaning of a *time series*. For A-theorists and tense theorists, the real time series is the A-series (or tense series). For B-theorists and tenseless theorists, the real time series is the B-series (or *tenseless series*).

There is also another important debate about time.[2]

2.2. Reality in Time

Some responses to McTaggart's arguments give rise to another debate. This debate is not directly about the reality of passage or change in the past, present, and future. It is about something more closely relevant to the subject of this book. This is a debate about the reality of things *in time*; one might also say that it is a debate about the reality of things *at times*.

The debate between presentists and eternalists concerns the reality of things—events, particulars, states, etc.—in any time other than a present time. The main two positions are *presentism* and *eternalism*.

2.2.1. Presentism

This is the position that only the present is real (or exists). The past and future are not real (or do not exist[3]). Only what is in the present exists or is real; anything at any other time than the present is non-existent or unreal.

Thus anything in particular (e.g., cats, dogs, a sandwich, the Big Bang, feelings of irritation, my potbelly, or that great haircut I had in the late '90s) that is only past or future (e.g., my haircut, the Big Bang) is not real. If it is present (e.g., my potbelly, that cat staring at me, half of the sandwich by my laptop), then it is real.

2.2.2. Eternalism

This is the position that no matter which times we call present, past, or future, anything in or at any time is real. It does not matter which time is present to the reality of things at all times.

Take any particular thing (e.g., cats, dogs, a sandwich, the Big Bang, feelings of irritation, my potbelly, or that great haircut I had in the late '90s). Even if these particular things are only past or future (or however else we might want to put it), it is still the case that they are real. Of course, if these are present (e.g., my potbelly, that cat staring at me, half of the sandwich by my laptop), then they are real too.

This distinction is important. Holding presentism, holding that only what is present is real, or the opposing position, eternalism, that what is at other times is real too, makes a distinction in how one understands the way the real structure of experience occurs in the world. It makes a distinction in how one models the constitution of experience, with significant implications—e.g., it makes a distinction in one's models of experience's constitution in space (as I discuss later).

If presentism is true, any real structure of experience must be present experience. However, if eternalism is true, the real structure of experience can be over multiple times. Indeed, as we see in the next chapter, if relativistic physics is true, we might want to say that the structure of experience *has* to be over multiple times.

2.3. Tense Theory and Tenseless Theory

One distinction already mentioned in the metaphysics of time is between two ways that we think of as the ordering of events in time. McTaggart describes the ordering of the events in terms of two series already introduced: the A-series and the B-series. To summarise the difference between these series, I present the following:

- *The A-series* is the series of events running from the far past, through the recent past, through the present, through the near future, to the far future. Events have different positions in the A-series—e.g., the Heat Death is in the far future, my writing this is present. Sometimes, philosophers call positions in this series *tenses* or *A-properties*.
- *The B-series* is the series of events ordered by B-relations: precedence, simultaneity, 'earlier than', 'later than', 'at the same time as'. No matter the moment in time, events stand in the same B-relations to each other, and thus in the same position in the B-series.

This distinction in two ways of ordering events in time serves as background to the A-theory and the B-theory (these theories get their names from these series). However, I need to make explicit two other distinctions between opposing positions. These distinctions are so closely related to the A-theory/B-theory distinction that they are often conflated with it. However, they are not identical.

The first distinction is the aforementioned distinction between *presentism* and *eternalism*. Debates between presentism and eternalism are often identified with debates between A-theory and B-theory. Yet some eternalists are A-theorists (e.g., Lowe 2003).

A slightly different set of theories around a debate is closer to the A-theory/B-theory debate; indeed, many theorists do not distinguish them. This is the debate about tense, and the opposing positions are the tense theory and tenseless theory. The debate here is whether or not tense is absolute, fundamental, or irreducible to anything else (tense being positions in the A-series) (Mellor 1998; Lowe 2003).

So far, this looks like the A-theory/B-theory debate. However, there is an important difference. A-theory may be identical to tense theory. However, the tenseless or detenser position is not necessarily identical to B-theory. Tenseless theorists may not endorse the reality, or at least the *absolute simpliciter* reality, of the relations that construct the B-series.

The issue is with the B-relations of simultaneity and succession. Given relativistic physics, many instances of simultaneity and succession are only relative (while others are absolute). With succession, the latter absolute relations are relevant to the definition of complex structures in time and space. However, with simultaneity, the latter absolute relations are not so relevant.

It is important to detail those differences here. It concerns complex structure, and, through that, it concerns the structure of experience. It has consequences for our models of how such experience gives rise to what we might typically identify as errors such as illusion and hallucination.

2.4. Relativity

For many, modern physical theory makes some counter-intuitive claims. However, when it comes to experience, one might think that these counter-intuitive claims do not matter.

First, these counter-intuitive physical claims do not matter to whether or not they are true. As evidenced by many current physical theories (such as quantum superposition or multiple universe theories), there is no 'intuition' constraint on physical theory.

Second, these counter-intuitive physical claims do not matter to understanding phenomena on the human scale. They only matter on scales (relative to humans) of the very small, the brief, the fast, the far away, the heavy, the big, or the long term. Human life (and especially human experience) can safely ignore it. We are not that small, brief, fast, far away, or big, and we do not last that long. Although we are physical beings, the claims of more intuitive[4] physics, such as Newtonian physics, are sufficient for accounts of human life. For example, they are sufficient for such things as action and technology: We can use engineering based on Newtonian principles to land on Mars. This more intuitive physics is enough to make most predictions on the human scale.

Although I think the first thought is correct, I think the second thought is not. Sometimes ignoring the weird physics can lead to difficulties in solving problems. For example, ignoring relativistic effects led to technical problems in the development of GPS (global positioning systems).

Furthermore, even if there were no problems in the everyday world, in any theory or account of what is really the case, counter-intuitive physical theories are the current physical accounts of how things are. One may explain something intuitively using Newtonian physics. However, if this account clashes with what is the case according to contemporary physics, then, if contemporary physics is correct, this account does not capture what is really going on.

When considering the actual structure of experience, I am primarily concerned with what is really going on. I am not primarily concerned with giving intuitive accounts of what is going on.

I argue that, when we think about the structure of experience in time, thinking about it from the perspective of relativistic physics involve issues and solutions different to ones given intuitive concepts of time. In particular, there is a difference in the role simultaneity plays in the structure of experience.

2.4.1. *Relativity in Physics*

Einstein's physical theories of space and time are often called theories of *relativity*. However, physical theories of relativity do not originate with Einstein, nor does a relativisation of what one might also think of as absolute. For example, Galileo formulated the principle of relativity, applying it to motion; motion is conceivably absolute (as it is under Newtonian physics).

The main point about relativity is simple: The definition of something (such as a property or relation) is relative to something; its values (such as measure or orientation) vary depending on different references. In principle, the reference—'something'—can be anything. For example, there could be property that is relative to human thought, moral truth, or even to God's perspective.

In contemporary physics, the relativity with which we are concerned is to arbitrary and conventional coordinate systems or *reference frames*. Reference frames or coordinate systems are ways of calculating values for spatial and temporal location and extent, with an origin defined by the frame. From this, we can define various physical properties, including location, speed, and direction.

'Arbitrary' and 'conventional' refer to the principle that no individual or subset of these frames are unique in defining the property *simpliciter*; there are no privileged frames. One can change the frame, getting a different set of values for that property, or different properties, for something, and yet those other properties are as real or genuine as those according to the first frame.

Think of *bigness* or *smallness*. If one is quite unreflective, then one might hold that there is such a thing as absolute and real *bigness* or *smallness*. However, it is far more correct to say that something being big or being small is relative to other things. Furthermore, one can vary what one picks to define bigness and smallness arbitrarily and conventionally. Perhaps, according to your sister's house, your partner's house is small; perhaps to your own house, it is big. Both are equally true. Beyond that, the question as to whether your house is *really* big or *really* small has little meaning.

When we say that a property is relative, we mean it is not absolute; it varies according to some frame or coordinate system. Yet this does not mean that we can say something does not *really* have that property. What we should say is that it does not absolutely have this property.

For example, that something is 'far away' depends on a frame of reference defined by a location in space. The only kind of answer to the question of whether or not something is *really far away* is in terms of it being far away to some point in space (and by some standard, but let's ignore that here). The sincere question 'independent of its position relative to some spatial locations, is it really far away or not?', makes no sense.

Similarly, in post-Newtonian physics, except for one significant exception (light), one can talk about speed and velocity as just as arbitrarily, conventionally, and relatively defined (I discuss this further later).

We can see this relativity by considering how we answer the following question: What is the real speed or velocity of my sister's car as she is speeding away with our shared inheritance? Any answer in contemporary physics is nonsense. To get a sensible answer, I should ask what the speed or velocity is of her car relative to me, standing there on the road with the fake documents, the road, the earth, and the police car coming the other way.

Yet one could deny that something really has a certain property even if it satisfies all the conditions of having it relative to something. If does not really have that property if it is necessary that the property be absolute. If the fact that something is not absolutely F means that it is not F, then that something is relatively F means that it cannot, in fact, be F. If it is in the concept of, say, three-dimensional shape that the shape be absolute, and not just that shape be relative to something else, then, if we held that it is that shape relative to one thing, but not that shape relative to something else, we should say that it is not really that shape.

This is why some philosophers argue that relativistic concepts of time are not really concepts of time. For example, Bergson insists that physical, relative simultaneity—what he calls 'learned simultaneity'—is a mathematical relation that is not simultaneity. "What ought you to do," he writes, "if you felt sorry for the poor philosopher condemned to tête-a-tête with reality, acquainted with it alone (rather than physical theory)? You would give another name to learned simultaneity, at least when you talk philosophy". (Bergson 1999, 46). It is not simultaneity because it is not absolute, and (he goes on to argue) simultaneity must be absolute.

2.4.2. Relativity and Subjectivity

Before moving on to the specifics of relativistic time, I need to highlight a distinction that is relevant to thinking about experience and time. This is the distinction between relativity and subjectivity. In a later chapter, I discuss what I mean by subjectivity (and *subjective activity*) in more detail. For now, I want to discuss it as a means of clarifying understanding of relativity.

Subjectivity is not the same as relativity. There can be relativity without any subject at all. For example, as discussed earlier, size is relative; Jupiter is larger than Earth; there are no subjects involved in this relation. We can also have relativity that involves subjects. For example, I am tall relative to my sister, short relative to my brother. These are relative sizes; they involve definition according to subjects (my sister, my brother, and myself).

Yet still these are *not* subjective properties; they are relative properties of subjects. There is nothing in either of us, or anyone else for that matter, that defines or generates my being tall relative to my sister. It does not depend on that either of us can think of a size relation that we stand in that size relation. As Jupiter is larger than Earth, I am taller than my sister is because of my physical, non-subjective properties. A non-subjective lump of clay that is the same size as me would be taller than my sister is; there is nothing subjective about this.

Similarly, my distance from something is dependent on my being somewhere, on having spatial properties, not on anything subjective. Again, I do not have to be a subject to give rise to that relative property. In defining such relative properties, there is no need to consider anything inside of me or dependent on me (what I shall later call *subjective activity*).

A special kind of relativity informs the convictions of tenseless theorists. In contemporary physics, this is the relativity of time (and especially tense).

2.4.3. The Relativity of Simultaneity

A common example in the literature about relativity theory is a thought experiment involving two observers, one on a moving train and the other on a platform (e.g., Dainton 2001; Balashov 2010; Power 2010).

Say you are on a train passing through a station: You can see the platform approaching, the board bearing the station's name moving by, the various people grabbing their bags and moving towards the edge of the platform, other people just standing and waiting.

Relative to you and your train, the station is moving. To yourself and the train, you are *not* moving (we assume for now, but I criticise this assumption later). In physical terms, you and the train occupy the same *inertial reference frame*. Anything that has the same velocity as you also occupies this inertial reference frame. However, nothing that moves at a different velocity shares your inertial frame.

Continuing the example, let us assume that your train is not decelerating, but is moving constantly through the station. Let us call the train's frame A. Call the frame of the station (and those standing still on it) frame B.

According to the train and your frame (A), the station frame (B) has a particular velocity: It moves at a particular speed in a particular direction (e.g., at five miles an hour to the left, Figure 2.1). According to B, to a person on the station platform, A, the train, is in motion in the opposite direction (to the right, Figure 2.2).

Therefore, for each of us (A and B), the others are in motion. However, for each of us, we ourselves are not in motion; we are at rest. Again, if several bodies are at rest to each other, they share the same inertial frame *by definition*. Yet to do so, they must have the same velocity: the same speed *and* direction. (This is important later).

2.4.3.1. Arbitrary Inertial Frames

Given the previous discussion, one might ask from which of these frames is genuine or real motion measured. Are the trains really moving or really at rest? If they are moving, are they really moving at the velocity measured by the platform? Might their real motion be as it is according to the trains or even by some meteor skirting the atmosphere?

If we had a geocentric model of the universe, we might easily answer this question by saying whatever is genuinely or really in motion is whatever is in motion relative to Earth. However, the geocentric model is false.

Newton defines the frame of absolute space as the frame of genuine or real motion; later theoreticians expand this in the theory of *the ether* (Lange 2002). However, with the theory of relativity, this changes. From Einstein onwards, we define motion arbitrarily and conventionally. The genuine or real motion depends on *which* frame we choose. When it comes to deciding something's velocity, there is *no* privileged frame. Thus there is no frame-independent absolute velocity (*ibid*).

This is important, as I discuss later, because another consequence of the theory of relativity is that time is *also* relative.

2.4.3.2. Relative Space and Time

According to the theory of relativity, along with the velocity, the distance between two objects or points and the duration between two events is different for different inertial frames. Most significantly, this applies to events that, according to some inertial frame, are *simultaneous and separated by a particular distance.*

This relative simultaneity and succession is because the speed of light in all inertial frames is invariant. The same light signal has a speed of (roughly) 3 × 108 metres per second no matter which frame you chose. As a result, the simultaneity and distance of events change.

We can see this in an example (illustrated by Figure 2.1).

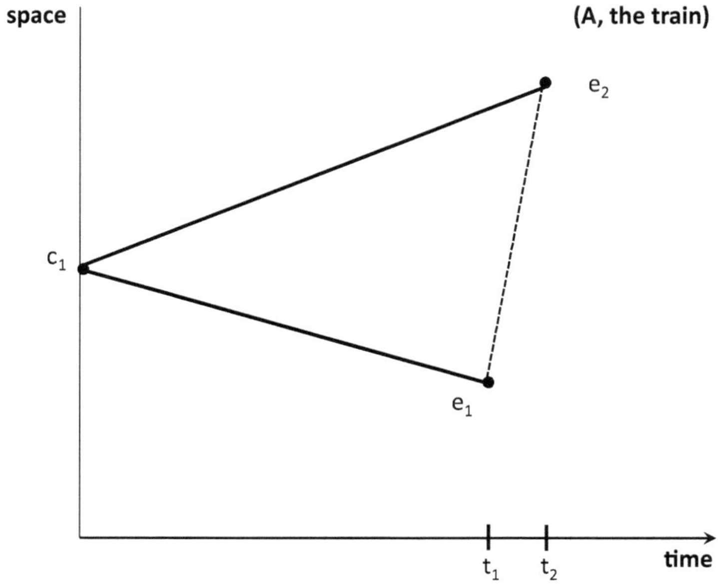

Figure 2.1 Relative succession

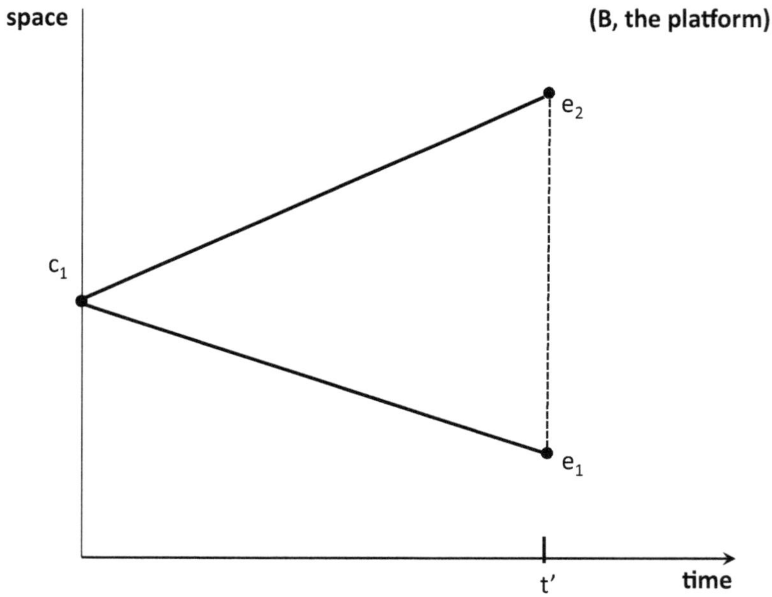

Figure 2.2 Relative simultaneity

Someone is standing exactly in the middle of the platform (B) as another goes by on a train (A). At each end of the platform is a perfectly reflecting mirror.

The person standing on the platform lights up their phone, causing a flash (c_1) that casts light in all directions. Two beams of light travel in opposite directions along the platform and each bounces off a mirror at either end (e_1 and e_2).

Figure 2.1 illustrates what happens for A. One end of the platform has moved closer since the flash, while the other has moved farther away, along with the mirrors.

Thus the distances travelled by each beam of light over that time are *different*. However, the speed of light is the *same* in both directions. Because of the difference in their distances, light takes *longer* to reach and strike one mirror (e_2) than it does to reach the other (e_1).

Figure 2.2 illustrates what happens for B. The mirrors are *equidistant* from the source throughout the passage of both beams of light. There is no change in the distance between person and either mirror between the time when the phone is lit up and the time it reflects in the mirror. This is because there is no motion between them; the lighter, mirrors, and platform are in the same frame. Moving at the same speed, then, each beam of light strikes its mirror (e_1 and e_2) at *the same time* (t'). The two events e_1 and e_2 are simultaneous.

Thus what is simultaneous on the platform is successive on the train. This holds for any two events separated in space, as the change in motion changes the distances between them and thus the amount of time it takes light to travel.

The relativity works both ways. It also applies to *succession*. Some events are successive according to only some inertial reference frames (e.g., A, the train).

2.4.3.3. Exceptions to Relativity

However, there can be absolute simultaneity between *spatially coincident* entities—entities that occupy exactly the same points in space. If coincident entities are simultaneous in any inertial frame, they are simultaneous in all inertial frames.

For many philosophers, the only absolute temporal relations between spatially separated entities are the *succession* of cause and effect (e.g., Mellor 1998). No causal relata are simultaneous in an inertial reference frame. As seen in the train example earlier, a causal chain can extend over space.

Thus space and time are not absolutes. In different inertial frames, different spatial distances and durations hold between the same two events. That is, the duration or distance between any two events depends on your *velocity*, and, for those moving at a different velocity, the duration and distance between these exact same events are different.

However, the theory of relativity also holds that the variations in space and time between frames are *connected*. In modern physics, we conceive of this connection as *aspects* of one four-dimensional extension: *Minkowski space-time* (e.g., Minkowski 1952; Balashov, *op.cit;*Sider 2003). Space-time is invariable across *all* frames. Space and time are not; duration and distance in space-time is similar to depth and width in space.

To finish, let us introduce some specific terminology:

i) *Space-time points* are *events*, analogous to pre-relativistic spatial locations and temporal moments.
ii) *Extensions between events* are *intervals*, analogous to pre-relativistic distances and durations.

Just like distance and duration, intervals are only extensions between events. One cannot assume that there is anything physical over them connecting the relevant events. Stating that two events are *connected* over an interval neither explains *how* they are connected or *what* connects them.

As already discussed, however, relativity does define some physical connection over intervals. In relativity, *causal relations* connect events. That is, what connects two events is that one event causes the other. We saw this in the example of the light striking the mirror. Indeed, we experience it every day in the physical events that occur around us.

What this means is that, so long as we can give frame-invariant or absolute physical connections between events, we only have defined connections *over* time. Strictly speaking, real structures are not states (or *static* structures). They are changes or processes. They are *dynamic* structures. That is, the physical connections currently defined under relativistic physics require that the only real structure is dynamic structure.[5]

What is dynamic structure?

2.5. Temporal Structure

The temporal structure of an entity is the collection of that entity's intrinsic temporal elements, properties, and relations. That structure comes in two important kinds:

- *Static Structure*: A static structure has the same elements, properties, and relations at each time of its existence.[6] A static structure that occupies multiple times has the same elements, properties, and relations at each time of its existence.

  Because its elements, properties, and relations are at a time, a static structure can occupy one time—i.e., it can be an instantaneous static structure.

Note this means the structure is in time, not outside it. It is mistaken to refer to, for example, mathematical entities, or other abstract *atemporal* entities as static.

In addition, I refer sometimes to a static structure as a *state* or an *unchanging* structure in time. One might also call it a *synchronic* structure.

- *Dynamic Structure*: A dynamic structure has different elements, properties, or relations at each time of its existence. A dynamic structure that occupies multiple times has different elements, properties, or relations at each time of its existence.

  Because its different elements, properties, and relations are at different times, either a dynamic structure cannot occupy one time, or it reduces to a static structure.

  That is, there is no dynamic structure at one time, or it is indistinguishable from a static structure.

  In addition, I refer sometimes to a dynamic structure as a *process* or a *change*.[7]

These definitions imply that static and dynamic structures are both in time. They are not outside it. Abstract atemporal entities such as universals or mathematical objects are not static.

If simultaneity between spatially separated things is relative to arbitrary and conventional reference frames, then any static structure constituted by elements which are spatially separated (i.e., spatially extended states) are also relative to arbitrary and conventional reference frames.[8]

There is another aspect of this definition of temporal structure: These elements, properties, and relations are *intrinsic*. I put aside what 'intrinsic' means in relation to structure until I discuss the structure of experience in the next chapter.

2.6. Other Positions in the Philosophy of Time

There are some other positions in the philosophy of time that I put aside here. There are debates about the possibility of an unreal future, even in cases that one accepts a real past. This gives rise to the 'growing block' position on time (and is somewhat related to the tense debate): The debate between substantivalism and relationalism, and continuous and discrete time, and debates over the conditions of persistence.[9]

I consider these other debates, with respect to experience, are either indiscriminable from the debates on which I focus or are irrelevant to thinking about experience. Briefly,

- We can treat *growing block theory* like eternalism with respect to the past and like presentism with respect to the future.[10]

- Debates over a possibly substantial or possibly continuous time concern what lies beyond experience.[11]

The persistence debate is generally applicable only to continuants (such as objects, which exist in time) and not events (such as changes, which happen or occur in time). I assume that experiences are not continuants but events.

2.7. Conclusion

This chapter introduced the different concepts of time: tense theory, presentism, eternalism, and tenseless theory. How these different concepts relate to understanding of the structure and errors of experience is the purpose of the rest of this book.

We also discussed the implications for time from relativistic physics. Relativistic physics is important because it provides a significant motivation for the less intuitive (or allegedly less intuitive) concepts of eternalism and tenseless theory, and provides an argument against the more intuitive (or allegedly more intuitive) concepts of presentism and tense theory. It is also important because it concerns the real, absolute, and relative *physical* structure of things.

It has implications for how one thinks about the relationship between the spatial and temporal structure of things. In particular, it has consequences for the spatial and temporal structure of perceptual experience.

I turn now to the structure of perceptual experience.

Notes

1. One could refer to these different conceptions of change as, respectively, *A-change* and *B-change* (as I do in my unpublished PhD thesis (University of Leeds, 2010)). As I do not focus on the A/B distinction, I do not find that terminology useful here.
2. Some philosophers may dispute how I refer to temporal passage here. Some tense theorists may deny passage and some tenseless theorists may accept it.

   Prior argues that temporal passage, conceived as change in events, is not real; there are only changes in things (Prior 1968, 1–14). Since the concept of passage here is change in events (or the present's relationship to events), one might take Prior to reject this concept. However, Prior's motivations are not those of tenseless theorists. Given how he presents his argument, his position is more like a presentist approach to thinking about things in time. Events undergo only if they do so by being past or future; nothing is past or future; nothing undergoes passage. (This could mean that presentism must also deny passage. However, this book does not presume presentism (except in its effect on thinking about experience), so I do not engage with this possibility).)

   Tenseless theorists (e.g., Oaklander 2012; Deng 2017) argue that tenseless can easily have a concept of temporal passage entirely in terms of tenseless change. That is, there is a sense of *temporal passage* that is just a kind of tenseless change. I take it that this is a further argument against the need for temporal passage as conceived by tense theorists. As I am a tenseless theorist, this supports my own position. However, for this book, I find it more convenient (and more common) for temporal passage to refer only to the change defended by tense theorists.

3. For the discussion in this book, I do not make a distinction between existence and reality. For the most part, I refer to the 'reality' (and related terms) rather than 'existence' (and related terms). I find it discursively easier and see no significant difference to the discussion in doing so.

   Some metaphysicians would find this naïve or misled. If so, then they may find interesting results by shifting it to only using one or the other of these terms and their related terms.
4. One might wonder how one physical theory can be 'more intuitive' than another physical theory. It sounds strange to think that there is anything like an 'intuitive physics'.

   I take it that, whatever one means by intuition, modern physical theory is commonly thought to be in more conflict with it than earlier physical theories, such as Newtonianism. It follows that Newtonianism is less in conflict with intuition; this is all I mean by it being a 'more intuitive' theory.
5. For an accessible mathematical introduction to relativistic physics, see Feynman (1971).
6. One might prefer 'occurrence', wanting to keep 'existence' for continuants, and denying that structure exhaustively defined determines the continuants. I see no problem with 'existence' here, but that might be a consequence of my biased tenseless views. If 'occurrence' is preferred, then one should read 'existence' in this section as just meaning 'occurrence'.
7. In the literature on time and consciousness, there are 'synchronic' and 'diachronic' relations—e.g., Dainton refers to synchronic and diachronic relations of *co-consciousness* (Dainton 2000). One might wonder if those 'static' and 'dynamic' can be aligned with 'synchronic' and 'diachronic'—i.e., I may equate static and dynamic structure to synchronic and diachronic structure, respectively. I see no problem with doing that but find 'static' and 'dynamic' sufficient here.
8. In the literature, one might come across references to the tenseless theory as the position that time is a 'static block' or to tense theory as the position that time is 'dynamic'. One might wonder how these definitions of 'static' and 'dynamic' relate to my use here of 'static' and dynamic.

   As this is not a book focused on the metaphysical or linguistic debates in the philosophy of time, there is no space to answer this. However, briefly, my thinking follows that of Smart, although his distinction is in terms of three dimensions and four dimensions (Smart 2001). I hold that this use of 'static' and 'dynamic' is mistaken. It rests on the following assumptions: (a) temporal passage is necessary for change and denying it denies change; (b) if something is dynamic, then it undergoes change; and (c) if something is not dynamic, then it is static. Therefore, as tenseless theory denies temporal passage, it denies change, denies anything is dynamic, and so asserts that anything that exists is static.

   However, being a tenseless theorist, I deny (a). Temporal passage is not necessary for change. (Also, I deny (c); something need not be static if it is not dynamic. It can be atemporal; it can have no temporal properties; however, that is a separate issue.)
9. In addition, in eternalist tense theory, there can be two different concepts of how a unique present relates to change and tense. I discuss this in the next section.
10. For this book, we can treat growing block theory like a combination of these theories. This does not mean that growing block theory *is* a combination of

these theories. Graeme Forbes, a growing block theorist, asserts (in personal conversation) that a proper understanding of growing block theory is not as a hybrid of eternalism and presentism. If I understand him correctly, the growing block theory holds that there is a part of time that is real, a part of time that is not real, and a constant change such that more events in time become real. The past, present, and future come from that more fundamental distinction.

11. Similarly, the difference between the moving spotlight and its alternative either lies beyond the phenomenology, or, if it does not, I do not know what the phenomenological difference is (nor have I come across an argument for or description of that difference).

3 The Structure of Perceptual Experience

This chapter considers the structure of perceptual experience. It is important to present at least a basic or general outline of such structure in order to consider the possible ways in which experience can be in error under different positions on time.

This chapter has the following sections:

First, on varieties of experience. Perceptual experiences do not encompass all varieties of experience. There are other kinds of experience. To discuss perceptual experience in time, we need to demarcate what is and is not perceptual experience.

Second, on naïve realism. I focus on naïve realism in this book, not because I think it is necessarily the best theory of perception but for three reasons. First, I argue that naïve realism has obvious ontological commitments through which the philosophy of time is relevant to it. This is not so obvious with other theories of perception. Second, naïve realism has important epistemological consequences; if true, one can, as it were, 'read off' experience facts about the actual world. Third, naïve realism is familiar to anyone working on perception.

Third, on the spatiotemporal structure of experience. As I stated in the introduction, I assume that experience happens in time and that it might have some temporal structure. Similarly, I assumed that, given physicalism, if there is experience, then experience happens in space, and so it can have some spatial structure. With those possibilities in place, I draw on recent thinking about the Extended Mind Hypothesis (EMH) to conceive of experience as embedded and extended. I then look at this conception of structured experience in terms of the assumed position about time in this book: tenseless theory.

Fourth, on the relations in experience. Some relations of experience are spatiotemporal. However, I also have reason to posit other kinds of relations in perceptual experience. I consider three possible kinds of relations in experiences (kinds that can overlap in their members): intrinsic relations, causal relations, and perceptual relations.

Fifth, on what I would call the 'hidden' structure of experience. The support for the structure posited here comes from perceptual *appear*ances or phenomenology. Yet some of the structure is what I call 'hidden' structure. It is not apparent or phenomenological. It is not shown, or obvious, in experience. The subject is not aware of it. Typically, we posit it because of how appearances fail to correspond to reality (i.e., erroneous experiences (which I discuss in the next chapter)). The (hidden) kinds discussed here are subjective and amodal structure, including how we might understand them.

Finally, I do not mean for this account of the structure of perceptual experience to be an exhaustive account. Furthermore, I do not mean it as an account of what is relevant to thinking about perceptual experience in light of different theories of time. However, from the very beginning, time and error play a role in picking out the structure of experience.[1]

3.1. Varieties of Experience

It is common in the philosophy of perception to treat 'experience' as identical to 'perceptual experience'. It is also common to discuss perceptual illusions when talking about errors of experience. This book is about perceptual experience and its errors. For most of it, I mean by 'experience' and 'illusion' perceptual experience and perceptual illusion.

However, perceptual experience and perceptual illusion are not the only kinds of possible experience and illusion.

3.1.1. Non-perceptual Experience

Generally, philosophers of time agree that what we seem to perceive seems to be present. When seeing a bouncing ball, the ball appears to be bouncing *now* (for discussion, see, e.g., Butterfield 1984; Le Poidevin 2007; Callender 2008). A perceptually apparent O is an apparently present O.

Importantly, if it is the case that what we seem to perceive seems present, then the perception of time can only refer to *one* understanding of time. It can only refer to *present* time. In all cases, *time perception* is *present-perception*. In this view, we cannot perceive a time that is a past time or future time.

Yet there is reason to resist identifying, generally, the content of *experience* with only present things. Some experiences do not seem to be of the present alone. Memory experience, for one thing, is not of apparently present things. In addition, deliberation over temporal experience must allow the possibility of including experiences of duration and temporal order. As such, there is reason to resist identifying the content of experience with only perceptual content (which is apparently of present things).

This does not mean that the only temporal property that we perceptually experience is presentness. In sympathy with some casual comments of Russell (Russell (1998), and somewhat less casual assertions by Oaklander (in Oaklander 1995), my own view (Power 2012) is that one might perceive some kinds of brief temporal order and duration. This is temporal order and duration contained within the period of changes that we perceive, a period many contemporary theorists of time perception call *the specious present* (see Chapter 6).

If this is right, then we can encompass some complex temporal features, and so temporal structure, in a present, and have those features perceptible. Yet some alleged temporal experiences (including temporal illusions) are of duration and temporal order that cannot fit within this perceptible present (or specious present). I discuss this issue in more depth in Chapter 9.

The immediate example is memory *experience* of particular past episodes. Generally, theorists accept that there is a phenomenology to some memories. There is something apparent to us when we remember. What is experienced in such remembering does not seem to be something present. The experiential content is in the past, at an earlier time to the act itself of remembering. In support of this suggestion, Husserl writes,

> I now remember vividly the terrain of a military exercise. I have the colour of the sky, the varying tints of the green of the meadows [. . .] These colours are past colours; namely, colours of objects that I am remembering but that certainly do not stand before me as objects that are now present along with all their determinations. [. . .] [T]he colours that are experienced there are ascribed to the past
>
> [. . .]
>
> In actual perception, the sensed colours are taken as belonging to the same temporal position as the perceived colours. The situation is analogous in the case of memory. The memorial appearance together with its body of sensuous contents (which themselves fall into the appearance) is taken to be the re-presentation of the earlier perceptual appearance; consequently, the sensed colour, just like the remembered colour, is taken as having been.
>
> (Husserl 2005, 243–244; see also MacKay 1945)

If we do have memory experiences, then this either one of two things:

- *Memory experiences are perceptual experiences*: We have perceptual experience of temporal features other than those that belong to (apparently) present things.
- *Not all experiences need be perceptual experiences*: Experiences need not be perceptions.

I suggest the latter: Experiences need not be perceptual experience. Perceptual experiences are as of apparently present things. Some experiences—e.g., memory experiences, are as of past events.

Thus, in general, I hold these positions:

- Experience corresponds to when there are appearances and/or to occasions in which something is apparent. When there is an experience, there is always *phenomenology* or appearances.
- That there is a phenomenology or appearances of something does not necessarily require that it appears to be *present*.
- 'Perception' or 'perceptual experience' is restricted to conscious episodes where something is apparent but necessarily appears to be present.

3.1.2. Non-perceptual Illusions

Considerations about non-perceptual experiences opens up the possibility that there are illusions (and other erroneous experiences) which are not perceptual illusions (or other erroneous perceptions). This is because, even if what we experience is not present, how it seems can still be mistaken. Depending on how it is wrong, an experience may be illusory, hallucinatory, or anosognosic (see later). I also think that an illusory, hallucinatory, or anosognosic experience can encompass a variety of experiences beyond merely *perceptual* experiences.

If there are memory experiences, things appear a certain way to a subject (as *past*). If that is right, then it is possible their appearances do not correspond to how those things are. This is a discrepancy between appearance and reality, independent of perceptual experience. As such, it satisfies conditions for being an erroneous experience. In addition, depending on precisely the way the discrepancy occurs, such erroneous experiences can be further specified in ways matching erroneous perceptions—e.g., if the discrepancy is in non-present properties, then the memory experience is an illusory experience.

Philosophers of perception understandably only consider perceptual illusions and hallucinations. Yet illusions crop up in discussions about memory and time as well. Notably, parallel with experience and time, talk about temporal *illusions* makes less sense in the context of conceiving of illusion as restricted to perceptual experience. If perceptual appearances are of present things, then, if illusions are only perceptual illusions, the appearances in a 'temporal illusion' could only concern present and temporal things. That is because illusion must concern *only perceptual* experience. If a perceptual experience includes the appearance of presentness, temporal illusion must concern only apparent temporal properties that are compatible with the appearance of presentness.

Again, one can have a temporal illusion—e.g., in memory—without it necessarily being that which seems to have the relevant temporal property also seeming to be present. As such, in general, I extend 'illusion' and 'hallucination' to apply to any variety of experience, not just to certain kinds of *perceptual* experiences.

These considerations are also the reasons for why I leave open the possibility of another idea: There can be illusions in *imagination*. One might imagine something and have an experience in that imagining. As such, there is *a way things appear to you* in that imagining. If the way things appear to you in imagining is mistaken, then one can call that an erroneous experience. As with memory experience, depending on what is mistaken, I can then categorise that erroneous experience as an illusory, hallucinatory, or anosognosic experience. For example, if the error is in an imagined property, then it is an imagining that is an illusion.

The point here is that this kind of erroneous experience in imagining is possible. It is not that there is any actual instances of such erroneous experiences. As of now, in the literature, there are no explicit cases of this kind of erroneous experience.

Yet I think we could interpret one actual case of mistaken experience as a case of erroneous imaginative experience. This is the *Perky* effect (Perky 1910; for discussion, Power 2011; MacPherson 2012b; Hopkins 2012; Nanay 2012).

Briefly, a simple description of the Perky effect is that a subject visually experiences something that is really there, but it appears to them as if they are visually imagining it. For example, experimenters asked subjects to visually imagine a banana on a screen before them and then to describe how they imagine the banana to be. In cases where nothing otherwise is projected on the screen, some subjects report an unpeeled banana. The experimenters then projected an image of a peeled banana on the screen at an illumination that made it visually detectable to the subjects. The subjects report visually imagining an image of a peeled banana on the screen.

There are a number of possible consequences to this. One possibility debated by Hopkins and Nanay is that the phenomenology of seeing is indistinguishable from visually imagining (Hopkins, *op.cit*; Nanay, *op. cit.*). However, this option is not open to me. I take it as a given that, phenomenologically, seeing is different to imagining.

I discuss another option in Power 2011. What appears to be visual imagining is visual perception—i.e., seeing. More precisely, what seems to be merely imaginary is in fact real. In that case, the content of the experience in imagining is mistaken. This is an erroneous experience with respect to its imaginary aspect. The subject of it is not imagining what is experienced, but it seems that they are. Depending on how we conceive of the phenomenological difference between imagination and perception, we might categorise the experience further as an illusion or hallucination. For example, if we hold that mind-dependence is a property, then this experience is an illusion of mind-dependence.[2]

Again, this is an example of how we might interpret actual instances of experiences as illusory imaginings. I can see various problems with it. For example, the idea that mind-dependence is a property that can be mistakenly apparent, and thus illusory, may feel a little forced (whatever worth such a feeling may have). In any case, this is not the place for a more thorough examination of this proposal.

This book concerns perceptual experience and time, and the errors involved. A reason that I am interested in perceptual experience over imaginative experience (or even memory experience) is that there is a particular importance to both the phenomenological character and epistemological role of perceptual experience. We are not motivated by the phenomenological character of imagination (or to a lesser extent memory) to hold that the contents of our experience are part of the actual world. We are also not motivated by the same kinds of experience that such content is part of the experience itself. For example, imaginative experience does not motivate us to believe that what we imagine is part of the structure of the world or of the process of imagining itself.

However, perceptual experience does provide such a motivation for holding that what we experience is both part of the actual world and of experience itself. In the rest of the chapter, I present what I consider the best reasons for thinking this and how it relates to the structure of experience itself. The main reason is the desire to hold that *naïve realism* is true.

3.2. Naïve Realism

Before introducing naïve realism, I must state what I mean by the *structure of experience.*

By structure of experience, I mean the collection (be it a grouping, aggregate, or unity) of intrinsic *elements (or parts)*, *properties*, and *relations* of experience. Sometimes, I refer to such a collection as the *constitution* of experience and to experience's elements, properties, and relations as its *constituents*.

I use 'intrinsic' in the Lewis and Langton sense: Experience's possession of intrinsic entities (properties, relations and elements) is compatible with both experience's being *accompanied* and experience's *being alone* (Lewis & Langton 1999). Here I assume that the structure of an experience is not anything extrinsic to it. It is not relative or dependent on something outside the experience itself. Still, an experience can be part of something else's structure. For example, if one holds that an experience is itself intrinsic to a complex process of beliefs and attitudes about the actual world, then the experience is part of the structure of that process. It does not follow from this that the process or the beliefs are part of the experience's structure. (Later in the chapter, I return to the concepts of intrinsic and extrinsic.)

One might wonder if one can say that an experience has a structure at all. One might take the position that experience is *simple*. For example, one might hold that (a) nothing complex (such as elements, properties, and relations) is intrinsic to experience or (b) that any structure we may be tempted to ascribe to experience actually belongs to something else, something related to experience, perhaps (e.g., physical events), but not experience itself.

If one is a physicalist about experience, then I think it is mistaken to hold (a) or (b). Even if one is not a physicalist about experience, I think it is mistaken to assume hold (a) or (b) for *perceptual* experience. This is because they require naïve realism to be false, and I think that is an assumption about something that is not yet settled.

What is naïve realism? A crude definition is this: It is the position that *appearances are reality*. It is a *naïve* position because it is that how things appear (whatever they may be) is how things are. It is a *realist* position because it concerns the appearance of real things.

For perceptual theory, those disposed favourably towards naïve realism provide more through definitions, with implications for perceptual experience's structure. They define naïve realism as the position that what we perceive partially constitutes our perception of it.

For example, Logue writes that we can characterize naïve realism as the position that "veridical perceptual experience fundamentally *consists* in the subject perceiving physical entities in her environment" (Logue 2011, 269, my italics). Nudds writes,

> [W]hen we introspect a visual experiential episode, it seems that we are related to some mind-independent object or feature that is present and is a part, or a constituent, of the experience. We can call this property—the property of having some mind-independent object or feature as a constituent—the naïve realist (NR) property of experiences. It is widely accepted that visual experiences seem to have the NR property; naïve realism is the view that some experiences—the veridical ones—actually do have it.
>
> (Nudds 2009)

If Nudds is right about perceptual experiences, then we can at least say that, if naïve realism is true, we perceptually experience a *part* (or *element* or constituent) of the experience. This part is the content of the experience, the content being part of the experience.[3]

Given another conception of perceptual experience, this account makes little sense. This is of perceptual experience as something that happens to us or that we *do*. In this sense, what we perceptually experience is not part of the experience—e.g., I seem to see a distant mountain; a distant mountain is neither something that I do or that happens to me.

This may make it seem as if we cannot conceive of perceptual experience in this latter way and have naïve realism be true. However, we can have both: What happens to us is an *element* or *part* of perceptual experience—a part different to the part that we perceive. If one talks of perception as something that happens to you (e.g., some set of neural events) and talks about perception as being partially composed by external things (e.g., a cloud moving over a mountain), then one is talking about different parts of perception.

Note that the perceptual experience seems only to be constituted by what is perceptually apparent.[4] It does not seem to be constituted by something happening to the perceiver (such as neural events), but does seem to be constituted by external things (such as shadows moving over distant mountains).[5] This does not require that we deny naïve realism. Naïve realism is not the position that what is perceptually experienced is the *only* part of the experience. It is the position that it is at least *one* part of the experience. The experience can have more parts, including something happening to the perceiver.

In any case, naïve realism means that experiences are not simple things. They have elements or parts. Their elements have properties, and there are relations between those elements. This structure is at least partly what is perceptually apparent (such as mountains). What is perceptually apparent (mountains) is complex, with multiple elements (snow, treeline, fields), properties (green and rocky), and relations (bigger than the hills nearby).

As I have stated, I am interested in how far I can take an account of perceptual experience as motivated by naïve realism. Throughout this book, I fall back often on naïve realism to motivate some of my claims and to weigh up claims about experience given different positions on time.

Yet even if one puts naïve realism aside, if a physicalist, I still think one should assume the structure of perceptual experience is in some way complex. In particular, it does not have the spatial and temporal structure of a simple thing.

3.3. The Spatial and Temporal Structure of Experience

To discuss the spatial and temporal structure of experience, it is necessary to introduce two general concepts. These concepts have specific analogues in particular debates about the mind. These are the general concepts of *embedded* and *extended*.

3.3.1. Extended and Embedded Experience

In general, for some entity X (e.g., a physical object) and some region Y (e.g., in time or space),

- X is *embedded* in Y if X occupies at least one point in Y's extent and does not occupy all points in Y's extent. For example, a single

instantaneous event is embedded in time by occupying one moment; a galaxy is embedded in space, as space extends beyond that galaxy.
- X is *extended* in Y if X occupies more than one point in Y's extent. For example, a process is extended in time if it has multiple stages, all at different times. Matter may extend throughout space if it occupies all points in space.

I assume that experiences are embedded in time. Whatever our metaphysical commitments, it is an uncontroversial fact that experiences happen *at some time*. They are not like abstract entities, such as mathematical numbers. Nor do they go on forever; they do not extend through the whole of time.

I assume that a condition of something's physicality is that it has spatial location. As such, I assume that, if one is a physicalist about experience, experience has spatial location. In addition, just as with time, experience does not extend throughout all of space. Thus experiences are embedded in space.

As such, I hold that experience is embedded in time and space. The next question is if experience *extends* through time and space.

This is more difficult to answer, especially with respect to time. I think that part of that difficulty comes from our temporal intuitions. I should say that not all of the difficulty comes from such intuitions. Some of it can come from other things, including intuitions about experience (especially once we abstract away from the phenomenology of perceptual experience).

For this book, let us return to naïve realism for the answer. As said, if one assumes naïve realism for perceptual experience, then what we experience appears to be a part of the experience. If this is right, then at least part of perceptual experience has the spatial and temporal properties of what is experienced.

There are three consequences of this relationship between naïve realism and the spatiotemporal properties of perceptual experience. The relationship includes the following:

(a) Constrains the spatiotemporal properties of perceptual experience. They must include the apparent spatial and temporal properties of what is experienced.
(b) Allows an investigation, through the phenomenology, into some (although not all) of the spatial and temporal properties of perceptual experience.
(c) Provides a means of undermining naïve realism. If no part of a perceptual experience can have the apparent spatiotemporal properties of the content, then naïve realism is false.

This last point is particularly important when we consider the kinds of erroneous experience there may be, and how they play a role in theories of perception. It seems that it should be easy to undermine naïve realism. However, this is not so easy once we consider how perceptual experience might *extend* through time.

3.3.2. Spatial Structure of Experience

I assume physicalism; if experiences exist, then experiences have spatial location. I also assume that experiences are spatially embedded. This raises a number of questions about any instance of experience:

(1) Where is the experience located in space?
(2) Is that instance of experience extended—i.e., located at more than one point in space?
(3) If the answer to two is yes, then through which region of space is the experience extended?

One way to think about answering these questions is through a recent debate about the mind. This is the debate about the spatial extent (or not) of cognition. This is debate about the EMH.[6]

EMH is that, in some cases, mental and cognitive processes are extended. Mental and cognitive processes are not merely *embedded*. They are not just located *within* a particular region of space. They extend *over* or *through* that particular region of space. The particular region in EMH is the region encapsulating the subject's brain and body, as well as the environment of that brain and body. Embedded mental and cognitive processes are located within the limits of the brain; extended mental and cognitive processes go beyond those limits to the world outside the brain. Rowlands writes,

> [A]t least some mental processes—not all, but some—extend into the cognizing organism's environment in that they are composed, partly (and, on the version I am going to defend, contingently), of actions, broadly construed, performed by that organism on the world around it [. . .] of manipulating, exploiting, and/or transforming external structures [. . .]
>
> [T]he function of the action [. . .] on these structures is to transform information that is merely present in the structures into information that is available to the organism and/or its subsequent processing operations.
>
> (Rowlands 2010, 58)

In addition,

> cognitive processes are often (and on some versions essentially) embedded in the environment. [. . .] [S]ome cognitive processes are

> dependent on environmental structures in the sense that these processes have been designed to function only in conjunction, or in tandem, with these structures. [. . .] But however tight we make [this] relation of dependence, it is still relation of dependence, not constitution.
>
> (*Ibid*, 68–69)

The debate about EMH is relevant in a number of ways.

First, this way of thinking about the differences between 'embedded' and 'extended' refers to the brain and body, and their environment. However, given how these concepts are distinguished, it is not obvious that they need to refer to the brain and body, and their environment. Furthermore, it is not obvious that they need to refer to embedment or extension through space alone. Given the comparison between time and space, there are reasons from the philosophy of time to hold that there are embedded and extended cognitive processes in *time*.

I have already introduced a more general distinction between 'embedded' and 'extended' mental and cognitive processes:

- *Embedded* mental and cognitive processes are located at some points in time or space but not at all points in time or space.
- *Extended* mental and cognitive processes are located at more than one point in time or space.

Second, I focus on experience and not mental or cognitive entities more generally. Therefore, I only want to know the answers if experiences are embedded or extended.

Even if we accept that experience is in some sense both extended and embedded, we can still understand and take a position on the EMH debate as applied to experience, applying it to experience; either experience is extended through the subject's environment (e.g., beyond the brain) or experience is only embedded in that environment (e.g., only in the brain).

If naïve realism is true, then it looks as if perceptual experience must extend into the environment. Part of the environment is what is experienced—e.g., a distant mountain—or at least its side. What is experienced is part of the experience; the experience has the side of the distant mountain as a constituent. Therefore, part of the environment is part of the experience; the side of the distant mountain is part of both the environment and the experience.

Third, one aspect of Rowlands's conception of an extended mind is incompatible with a naïvely motivated concept of an extended experience. Similar to Rowlands's extended mind, an extended experience extends into the experiencing subject's environment. However, a naïve conception of perceptual experience is not only composed of the "actions [. . .] on the world [. . .] of manipulating, exploiting, and/or transforming external [things]." The perceptual experience may be partially composed of these actions but it is also composed of external things, such as mountainsides and stars, things that are not actions nor things that actions must exploit. As far as the appearances are concerned, we can look at the stars without having to use those stars or their appearances in some way.[7]

Yet is it possible for our perceptual experience to include such external things? That question requires thinking about time. Just like space, time determines what can constitute experience in the physical world. This includes the brain and the environment in which it is embedded.

Fourth, it is important to stress that the difference between an extended mental entity and embedded mental entity in EMH is not merely the content. The debate around the EMH does not concern what mental activity is *about*. Instead, the debate is about the *constitution* of mental activity. It concerns the structure and constitution of mental activity.

Similarly, with respect to extended or embedded conceptions of experience, I am concerned with the *constitution* and *structure* of the experience.

3.3.3. Temporal Structure of Experience

Now that we have briefly discussed the structure of perceptual experience, we turn to the temporal structure. We might wonder if experience is temporally extended or temporally embedded.

Taking the original questions for space and adapting them for time, we get the following:

(1) Where is the experience located in time?
(2) Is that instance of experience extended—i.e., located at more than one point in time?
(3) If the answer to two is yes, then through which period of time is the experience extended?

Unlike the spatial analogues of this question, the answer depends on the philosophical debates about time.[8] We have already covered the different debates in the philosophy of time. When coupled with the conception of experience as spatially *extended*, the philosophical positions about time assumed in this book require that experience meet certain conditions regarding its structure.

As discussed earlier, throughout this book, I assume the truth of tenseless theory as motivated by relativistic physics. I assume there is relative simultaneity between spatially separated things. I do not assume there is absolute simultaneity between spatially separated things, nor do I assume tense theory or presentism. I discuss those only because of how they influence thinking about the structure of experience.

In my 2010, I argued that, if relativity theory is true, then one ought to hold the following is likely as follows:

(a) Experiences extend in time.
(b) Experiences' elements (or parts)[9] are distributed over time.
(c) In being distributed this way, these constituents are connected *over* time, not *at a* time. An experience's elements are distributed over different times—i.e., they occupy different times to one other. This is partly because the relations that hold between them are not relations obtaining between simultaneous things—i.e., at one time.

Why are experiences connected over time and not at a time? It is plausible that experience has spatially distributed elements. This is the case even if experience is something that happens solely within the brain—i.e., experience is *embedded* in the brain; even in such a case, experience extends over spatially separated physical activity (such as neural firings at different locations).

In my 2010, I argue that these elements must be connected over time because the elements are causally related (Power 2010). This is because of the following:

(a) In relativistic physics, variant temporal relations are arbitrary and conventionally defined.
(b) The elements of experience are not connected by arbitrary and conventionally defined relations.
(c) (From a and b) The only temporal relations between the constituents of an experience are invariant temporal relations.
(d) The only invariant, non-arbitrary, and non-conventional temporal relations to hold between things at different spatial locations are those of causal relations.
(e) Experiences have elements at different spatial locations.
(f) (From c, d, and e) The only temporal relations to hold between the elements of experience are those of causal relations.
(g) Causal relations have non-simultaneous relata (at least, in contemporary physical theory), and so their temporal relations are not instances of simultaneity.
(h) (From f and g) The only temporal relations connecting the elements of experience are not instances of simultaneity.

I have somewhat changed my mind since the 2010 paper. In 2010, I assume that the analysis in relativistic terms is sufficiently captured by thinking only about temporal relations. However, given Minkowski's space-time, I should think about relations over *space-time*—i.e., the relations that occur over *intervals* between entities at *space-time points*. These intervals are invariant, non-arbitrary, and non-conventional.

With space-time, this argument does not deny that some elements may be connected to each other at one time. The connection can be over an invariant spatiotemporal relation that, in some reference frames, is an instance of simultaneity.

Furthermore, even if causal relations, strictly speaking, connect the constituents, it is not that the constituents *cannot* be simultaneous. Although connected *over* time, one could conceive of a structure where two elements occur at the same time according to some reference frame. Yet causation does indirectly connect them through a third element—e.g., two events, along with a mutual effect as a third element, may constitute one experience. Drawing on the example in the previous chapter, for the light bouncing off the two mirrors, we can say the following:
Assume that

(1) I am in a position such that I see the light hit both mirrors.
(2) I occupy one frame of reference (e.g., the platform or the train), and both events are simultaneous in that frame. (As I argue in my 2010, such a 'single frame' for a subject is problematic. However, ignore that for the sake of this argument.)
(3) (In line with naïve realism) my perceptual experience extends far enough into space as to include the perceptual contents. That is, since I perceptually experience the light hitting both mirrors, the light beams hitting each mirror are constituents of my experience of them.

Then,

(4) My experience includes two events that are simultaneous according to my own frame of reference.
(5) This is still a constituted experience in line with the claims here because of the following: Both events—the light hitting the mirrors—are part of the same experience.
(6) As such, the experience has spatially distributed simultaneous elements.

However, even if we consider space-time intervals, this reasoning does not mean that these elements of the experience are *connected* at a time (and so simultaneous). The spatially separated constituents of this experience are only arbitrarily and conventionally, only *accidentally*, simultaneous. What relates or connects the constituents is not arbitrary or conventional. As such, it does not depend on their simultaneity. In my 2010, it is a causal link—e.g., in the example, through the passage of the light. Yet, even if it is something else, it is something that does not depend on the simultaneity.

It cannot depend on the simultaneity unless it depends on something arbitrary. In the example earlier, there is some frame of reference from which two (arbitrarily) simultaneous elements in another frame are (arbitrarily) not simultaneous. (This is the case even if the difference between simultaneity and non-simultaneity is too small to be detectable through human measurement.)

What are the consequences of this argument for the general temporal structure of experience? Of most importance is the following:

- Elements at different times can constitute the same experience; any relations between such elements are between things at different times; any properties had by such elements are possessed at different times.
- The first consequence may be obvious: That things happen at different times does not disqualify them from being elements or possessing properties or relations of a single experience.

In other words, that things happen at different times does not matter to their being constituents of the same experience. So long as they are connected in some way in space or time, they are eligible to be constituents of the same experience.

If this is right, then it has implications for one's thinking about how experience is embedded and extended in space and time. It has implications for interpretations of experience as erroneous as well as for how precisely an experience is erroneous.

3.4. Relations in Experience

Another idea important to this book is that of the relations relevant to experience. With respect to the structure and constitution (and thus the ontological commitments of experience), I consider the following groupings of relations to be the most important:

- Intrinsic/Extrinsic Relations
- Causal Relations
- The 'Perceptual' Relation

These groupings of relations help define, constrain, or open out experience's spatial (and thus physical) and temporal structure.

3.4.1. Intrinsic and Extrinsic Relations

We can conceive of an experience as having *extrinsic* or *intrinsic* relations.

(a) *Intrinsic Relations*: Intrinsic relations concern the structure of experience. They are any relations between the following:

- Constituents of experience
- Constituents and the experience
- Experience and itself (i.e., reflexive relations)

These are intrinsic relations because (a) one needs either the experience or constituents of the experience to be the relata and (b) as with intrinsics in general (according to Lewis & Langton 1999), the holding of the relation is consistent with the experience or its constituents *being accompanied* or *being alone*. No matter what exists or occurs other than the experience and its constituents, the experience has its intrinsic relations.

(b) *Extrinsic Relations*: Extrinsic relations concern the relationship between an experience—and between any of its elements (or properties)—and anything that is not part of the structure of the experience—e.g., inanimate objects in the world, other experiences (perhaps other experiences of the subject), and/or whatever else is held to occur separate to the experience itself. These are extrinsic relations because their relata must include the experience (or one of its constituents) and something other than the experience (or a constituent). They are relations that hold for an experience or between its constituents that depend on if the experience is accompanied or it is alone.

As an example of intrinsic and extrinsic relations, consider an apple, its shape, its volume, and its constituents (e.g., pips, skin, flesh, and stem). If one holds that the apple has its shape and volume intrinsically, one holds that it has this shape and volume intrinsically; it has these properties if other things surround it or if it is floating in an otherwise empty void. In contrast, its distance from something else—e.g., another apple—is not an intrinsic property of it, because it depends at least on something else.

Similarly, intrinsic relations of an apple are the distance between two points on its surface or the relative size of some of its different parts (e.g., the pips, the skin, the flesh, the stem). However, its distance from other things that are not part of the apple is not an intrinsic relation, but an extrinsic relation.

Note that what is intrinsic to an experience need not be intrinsic to its constituents. The relative size of pip and stem is not an intrinsic relation to either pip or stem; it depends for each on there being the other as well—e.g., the pip needs to be accompanied by the stem, the stem accompanied by the pip. However, this relation is intrinsic to the apple; the apple has this relation between its constituents whether the apple is accompanied or alone.

Going back to experience: Some relations and constituents may be necessary for experience and some may not. This is true of both intrinsic and extrinsic relations. For this book, the most important of either extrinsic or intrinsic relations are any that might obtain over space and time (if there are any).

We can see that by looking back at earlier discussions. There is a connection between this discussion about intrinsic and extrinsic relations and the earlier discussion about experience as embedded or extended.

- An experience E and constituents of E, for some region W in which the experience is *embedded*, can only have extrinsic relations to (a) W and/or (b) any constituents of W that are not E or E's constituents. (Yet these relations between E and other constituents of W can be intrinsic to W).
- An experience E can have intrinsic relations between that over which it *extends*. These need not be intrinsic relations of E's constituents.

The intrinsic/extrinsic distinction is related to *constitution* and to *structure*. Intrinsic relations partake in the structure of experience. They are also the relations in the structure—i.e., between the elements, properties, and relations of the experience. (In addition, I see no particular reason to avoid considering them as some kind of constituents of the experience, as well as relating its constituents, of course).

This distinction between intrinsic and extrinsic is also relevant to thinking about two important types of relations for the structure of experience. These types of relations, although not the only relations that an experience might have, are particularly relevant to discussions about the spatial and temporal structure of experience. In particular, they play an important role in working out the precise way in which certain experiences are erroneous.

These two types of relations are *causal* relations and *perceptual* relations.[10]

3.4.2. Causal Relations

By 'causal relations' I mean different instances of the causal relation in different sets of causes and effects. For example, there is an instance of the causal relation between throwing a ball at a window (a cause) and breaking the window with a ball (an effect). If mental events can cause physical events, then there is an instance of the causal relation between deciding to break the window (a mental event) and (via a causal chain including throwing the ball) breaking the window (a physical event).

I assume it is intuitive to think experience stands in causal relations to other things—e.g., as causing or being caused by something else. For example, the shattering window (a physical cause) causes my experience of exhilaration and fear (an effect that is an experience). My fear (a cause that is an experience) prompts me to run away (a physical effect).

A causal relation between an experience, its causes, and its effects is an *extrinsic* relation to the experience; the relation cannot be obtained without an accompanying cause or effect; neither the cause nor the effect are part of the experience. Although less intuitive, we want also for experience itself to have intrinsic causal relations.

Here are two ways to describe things in reference to causation:

- *Self-Caused*: Things that cause themselves to happen.
- *Other-Caused*: Things that other things cause to happen.

I assume that an experience is not self-causing; I assume that it is caused by other things (and can itself cause other things to happen—i.e., I reject epiphenomenalism). Does this lack of self-causing mean that experience only stands in extrinsic causal relations? Not necessarily.

If an experience is considered *as a whole*—i.e., not just a subset of its constituents, then the causal relations of that experience must be *extrinsic* relations.

However, the extrinsic relationship need not be the only relationship between causal relations and experience. There is also the possibility of a causal relation between *constituents* of the same experience. As with all such relations between the constituents, such a causal relation is an *intrinsic* relation of experience.

The possibility of intrinsic causal relations in experience is important. If, for whatever reason, we deny the possibility of such relations, then we set a condition on the possible structure of experience. In so doing, we also set a condition on the possible ways in which a particular experience is erroneous or veridical.

For example, in anticipation of later chapters about perceived space, the possibility of intrinsic causal relations in perceptual experience enables a perceptual relation between something in the subject and particular physical things. When we have perceptual experiences, we seem to stand in such a perceptual relation. Denying this possibility undermines these appearances.

3.4.3. *The Perceptual Relation*

Another possible relation considered in discussions about perception and perceptual experience is perception *itself*. In having a perceptual experience, the subject stands in a particular relation to what they perceive. This is a perceptual relation between the subject and the perceived object.

For some theorists, this is apparent in the experience itself. Some even use it to define naïve realism. In an extended quote, Logue writes,

> Naïve Realism holds that veridical experience fundamentally consists in the subject bearing the perceptual relation to things in her environment (and some of their properties). For example, according to the Naïve Realist, the veridical experience I'm currently having of the yellow, crescent-shaped banana on my desk fundamentally consists in my bearing a perceptual relation to the banana, as well as its yellowness and its crescent shape. [. . .]
>
> The "fundamentally" in the statement of Naïve Realism is important, as without it the claim on the table would simply be that veridical experience involves perceiving things in my (mind-independent, material) environment—which everyone but the idealist accepts. Naïve realism is a much stronger claim. It is a claim about the

> metaphysical structure of veridical experience, i.e., about the ultimate psychological facts in virtue of which facts about experience obtain.
>
> (Logue 2015, 207–208)

Leddington writes,

> [N]aïve realism is the view that to perceive is to stand in an unanalyzable, nonrepresentational, yet cognitively significant relationship to objects and events such as tomatoes and baseball games.
>
> (Leddington 2009, 492; for other discussions about the perceptual relation, see Crane in Gendler & Hawthorne 2006)

Is there such an apparent perceptual relation, such as that between a subject and the banana on their desk? One (naïve) way of thinking about it is this: I have beliefs and experiences of several things, including particles, unicorns, and bananas. Some of those things are like this banana I see here. I do not just believe in them or experience them. I perceive them. My perception seems to be a special relation between myself and such things, in that it does not relate me to the other things I believe in or experience.

If it is the naïve position that there is a perceptual relation, then such a position holds that a perceptual relation is apparent to us in some way. The relation is not hidden from us; it is not something we infer.

Yet what kind of relation is this perceptual relation? Leddington states that, for naïve realists, this relation is unanalysable and Logue that it is fundamental. As such, it is not something we are supposed to explain in terms of other things.

However, perhaps it is only fundamental relative to the concepts used in understanding perceptual experience. Perhaps we can understand it as a relation that is shared with other kinds of experiences.

So far, reference to experience has always been as an experience of *something*. I have not simply said that a subject 'has' an experience and left it at that. This is so even where there is a question as to the reality of what I experience. There is still a 'what' that I experience—e.g., the unicorn when I visually imagine a unicorn.

We might conceive this as any experience as being a particular kind of relation—e.g., the relation between a subject and the content of experience, a relation that is particular to experiences. It is absent in other kinds of attitudes (e.g., an unconscious belief in unicorns) or other kinds of relations in which subjects stand (e.g., the relative height of a person and a horse).

However, in cases of experience as found in imagining the 'something' is not real. The imagined unicorn is merely an *intentional* entity. This kind of experience is like belief or hope, in that what is experienced need not be real (for discussion, e.g., Sartre 1986).

This raises a common problem with conceiving of intentionality as a relation. This is the problem of understanding how there can be a relation in which one relatum is unreal (e.g., Brentano 1874; Jacquette 1985; Jacob 2014).

Although there may be such an intentional relation in perceptual experience, there need not be such a relation. Unlike other kinds of experience, there is no problem of unreal relata for a relation in *veridical* perceptual experience. A perceptually apparent thing is an apparently *real* thing. With veridical perceptual experience, what is experienced is real. As such, for perceptual experience, we might hold that it is or involves a special *perceptual* relation, and this relation is not an intentional relation. Unlike imagination, it is an unproblematic relation to something real.[11]

However, such a perceptual relation has problems with particular erroneous perceptual experiences in which what we experience appears to be real, but is, in actuality, not real. For example, in cases of illusion and hallucination, the problem is that subjects seem to be perceptually related to something (with illusion, a property; with hallucination, an object) through the experience. Yet since it is an illusion or hallucination, there is no such thing. So what are we to say about this 'perceptual' relation? Is it not simply a version of the intentional relation?

Some may respond to this by asserting that, if the subject seems to be experiencing something that it turns out that they cannot experience, then they experience something else instead. This gives rise to a class of perceptual theories that involve commitment to *indirect realism* or *indirect perception*. We experience much of what we think we experience indirectly via something else.

For example, a typical sense-data theory holds that we are aware of objects that are external (meaning, outside our minds or brains) *indirectly* via something internal—*sense data* in our minds or brains (e.g., Jackson 1977; Hilbert 2004; for recent discussion, Logue 2015).

Others may try another approach, denying the need for a relation at all. In an experience, that the subject is related to something is a *merely possible* condition of perception. One can have a perceptual experience without there being *anything* perceived (e.g., adverbialists such as Tye 1975; for discussion, Jackson 1975).

As such, one may deny that there is a need for an intrinsic relation in perceptual experience. Alternatively, if there is such a relation, it is to something mind-dependent and not to anything external.

Such answers seem to me to hold that the intrinsic structure of imaginative experience sufficiently captures the intrinsic structure of perceptual experience. For perceptual experience, it is enough that it has the structure of experience had in imagining—e.g., it has intentional content and any 'relation' is an intentional relation; it has Brentano's 'mark of the mental' that holds between a subject and merely thought or imagined things.

Nevertheless, the structure in imagination clashes with the phenomenology of perceptual experience. As such, I want to resist this account of perceptual experience.

Yet, by definition, an illusion or hallucination is a perceptual experience in which, despite appearing real, what we experience is not real. At least regarding the illusory or hallucinatory entity, we seem to be perceptually related to an unreal thing. If we are related to an unreal thing in perceptual experience, we are no more compelled to commit to a perceptual relation than we are to a relation in imagination.

This undermines the need for a perceptual relation at all. If the phenomenology does not compel us in illusory or hallucinatory cases to hold that we stand in a relation to something real, then why does the phenomenology compel us in the veridical cases to hold that we stand a relation to something real?

This suggests another response, one that may appear more difficult to maintain. We resist the interpretation of erroneous perceptual experiences as being illusory or hallucinatory. We try to make sense of the errors in experience without involving cases where what appears to be real is not real.

To do that, we need to consider the structure of erroneous experiences as they occur in space and time. Through thinking of their structure this way, we think about that structure under different philosophical positions of time.

3.5. Hidden Structure in Experience

This final section concerns three possible features of experience that descriptions and explanations of perceptual experience may include. These features all have one thing in common: They are generally not consider perceptually apparent. They are not part of the phenomenology. Instead, they are elements of the experiential structure that some claim lie under or explain experience, including its phenomenology. As such, I call these *hidden* features of perceptual experience. They are as follows:

(i) Subjective elements of perceptual experience
(ii) Imaginative elements of perceptual experience
(iii) Representational elements of perceptual experience

I do not think these features are central to this book. This is primarily about the trustworthiness of perceptual experience in light of different phenomenological descriptions and ontological commitments. These hidden elements are what we get when we have settled such descriptions and commitments. They come out of various kinds of erroneous perceptual experience, out of ways in which appearances and reality diverge.

The motivations for positing them come into the debate after the discussion of this book is settled. Once we have settled on an ontology, and an acceptable phenomenology, the kinds of erroneous experiences motivate positing these features. For example, if we end up with a perceptual experience that is a hallucination, we could hold that the perceptual experience has a hidden feature that is subjective or amodal or representational.

This subjective feature is sufficient for perceptual experiences, needing nothing objective. However, presumably perceptual appearances are at least of objective things; thus, trustworthy perceptual experiences need a dependency on something objective. Thus the sufficiency of this subjectivity conflicts with perceptual appearances.

Yet once we are so committed, they can play a role in explaining such error. Thus before turning to error, it is important to outline them.

3.5.1. *Subjective Activity*

In Chapter 2, I discuss subjectivity in light of relativity. Here there are two important senses of 'subjective' for the structure of perceptual experience.

In a work such as this, where the concern is with the real structure of things, and part of the discussion concerns the difference between embedded and extended, I take 'subjective' to mean the following: Something is subjective if it is something that *depends on subjects*. If there are no subjects, then there is no subjectivity; if there is subjectivity, then there is at least one subject. In stating this, I do not attempt to distinguish what a subject (e.g., a person) is from anything else (e.g., a car the person is driving).

However, I do assume that subjects have physical aspects (e.g., brains and bodies) and these physical aspects are embedded and extend in space. Furthermore, I assume that there are processes or products of processes that are *within* these subjects—e.g., neural activity in the brain (or, if we accept that bodies are part of the subject, respiration). I also include mental processes and states—e.g., beliefs, thoughts, desires, or deliberations.

I call such processes or products *subjective activity*.

Here 'activity' is only means something that happens—i.e., it is synonymous with 'events'. I do not mean by it things that subjects or agents *intend* to do (as perhaps some philosophers would have as conditions for *action*). Instead, it refers to things that subjects *do* or that *happens in them*, intentional or otherwise.

In the context of experience, subjective activity includes the neural activity identical to or correlating with the mental activity associated with an experience. We might not only associate this subjective activity with experience. Subjective activity can be one constituent among many of experience (for example, if experience extends beyond neural activity) or even encompass the whole structure of experience (if experience is embedded in neural activity).

I make the following assumptions about subjective activity. The first is that subjective activity is something that only happens within a subject. This might sound strange when applied to any act or action whatsoever. Driving a car is something a subject *does*, so is it not a subjective activity? Yet it is something that is at least partly outside you (e.g., the parts of the car you move). The changing parts of the driven car are part of the activity, but lie outside the subject.

There is some leeway in calling this subjective activity. However, to discuss perceptual experience, we need to distinguish what happens within the subject and outside the subject, even in such activities as driving a car. 'Activity' alone does not pick out what happens within the subject. Car driving is a counter-example to that; such activity is constituted by events outside the subject. The movements of the gears, the pedals, the pistons, etc., may be part of an activity that a subject does or at least causes to happen (e.g., through pressing the accelerator or turning the key). Yet it is also outside them as something that happens in the car.

To distinguish activity and activity within the subject, I refer to an activity that is wholly within a subject as subjective activity. Subjective activity comes from or is within the subject. If we remove the subject, there is no such activity.

Before moving on, I need to address another possible understanding of 'subjective': Something is subjective by being only the content of a subject's mental state—i.e., it is *merely intentional* content. Typically conceived, intentional content includes whatever one's mental state is about—e.g., the intentional content of my idle recollection is an excellent meal I had in Paris; the intentional content of my hunger is food.[12]

However, I do not use 'subjective' in this way. On the one hand, this understanding is too restrictive, being only about mental states. I do not presume that all there is to a subject or subjectivity is their collection of mental states. A subject can be embodied and so can have non-mental bodily aspects or parts. A subject's brain and nervous system can have non-mental neural activity (e.g., processes involved in motor action or digestion).

On the other hand, ambiguity about this meaning of 'subjective' makes it too broad. It is common to refer to intentional content as what a mental state is *about*. We can think about many things, many that we call subjective and many that we call objective. Intentional content can be subjective (e.g., reflection upon one's own moods or bodily state) or objective (e.g., a belief about the speed of an object relative to the moon).

As said, how I use 'subjective' might not be the same as others use it. However, even given the general metaphysical realism of this work, I need to talk about what happens in the subject of experience. I use 'subjective' to refer to that.

3.5.2. Imagination in Amodality

Another important feature of perceptual experience is the sense that many things that we perceptually experience also appear somehow to go beyond what is perceptually experienced. There are two ways to understand this.

First, one has a complex of experiences, only one of which is a perceptual experience—e.g., one has a perceptual experience accompanied by a memory experience and an imaginative experience; one sees a horse, recalls seeing on an episode of *Black Beauty*, and imagines Black Beauty as a unicorn. In having this complex experience, there seems to be more than what is perceptually experienced. As such, perceptual experience itself does not exhaust what is apparent; it is only a part of the phenomenology.

Second, specific to the perceptual experience, independent of other kinds of experience, perceptual experience in some way appears limited. Perceptual experience of the world is not and does not seem to be of the entire world. Something lies beyond what is perceptually experienced. The experience seems to leave something out, something that belongs to what is perceptually experienced (and so we do not merely remember or imagine it).

Many perceptual theorists think that perceptual experience always comes with such an accompanying experience of a limit. There is a 'presence in absence' or *amodal perception*. Noë writes,

> Look at an apple on the table in front of you. You have an experience of a voluminous whole. In fact, you can only see the facing surface of the apple. It doesn't seem to you that you actually see the whole of it. After all, what could be more salient than the fact that you only see its visible features, its facing side? And yet, in seeing what is visible, you have a sense—a visual sense—of the presence of the apple as a whole.
>
> Similar considerations go for a cat, whose body is hidden by an obstacle, but whose head and tail show. You can visually experience the head and tail as belonging to the same cat, even though you can't actually see the part of the cat to which they are joined. Here it doesn't seem to you, erroneously, as if you see something you don't. Rather, it seems to you as if you are aware, in a perceptual modality, of something that is plainly out of view.
>
> (Noë 2005, 245)

Similarly, referring to the process underlying this, Breckon and Fisher refer to the way in which our sense of the whole of something is given by what is not just apparent as 'amodal completion', meaning:

> *Amodal completion:* the perception of a background object/shape continuing behind a foreground occluder despite the fact that one

> cannot see it directly. The occluded portion is invisible yet perceived by the viewer.
>
> (Breckon & Fisher 2005, 501)

That is, perceptual experience is accompanied by a phenomenology of the perception being *limited* or *partial*. There are a number of consequences for this in thinking about the structure of experience, especially as motivated by phenomenology through naïve realism.

When one naïvely claims that what one appears to see is what one sees, it is not an objection to the position that one sees only some of the object—e.g., I do see the surface and colours of the tomato, but I do not see the inside of the tomato, or the gaps between its atoms. It is no objection to naïve realism that, when a naïve realist claims that what is perceptually experienced is a constituent of the perception, the constituent cannot be the whole of the perceptually experienced thing. For naïve realism, a constituent can be part of what is experienced. To put it another way: Given naïve realism, there can be more to what is perceptually experienced than what is perceptually apparent.

As such, when it comes to a discrepancy between appearance and reality, we need to be clear about the ways in which that discrepancy can occur. If there is an experience of a limit to our perceptual experience, then there is not a discrepancy if reality goes beyond (or transcends) perceptual experience itself. If what we perceive (such as a tomato) does transcend perceptual experience, then, since this is how it appears (there is more to the tomato than what is perceptually apparent), it is not a case of erroneous experience.

However, we can still have some kind of erroneous experience around amodality. What we perceive (a) may not go beyond perceptual experience or (b) may not go beyond perceptual experience the way it appears to. Say that what we see seems to be one thing (e.g., a whole tomato) that goes beyond what we see of it (this side of the tomato). If it is in fact something else (e.g., a painting of a tomato), then, given amodality, there is a discrepancy between appearance and reality. There appears to be one thing (a tomato), but there is in fact something else (a painting).

However, one might raise a question about this analysis of perceptual experience. When we say the limit is *apparent* or the perceived object is *experienced* as transcending what is perceived, what do we mean by 'appears' or 'experienced'? What we cannot mean is that we have a perceptual experience of the limit or transcendent aspects of this. That would be the incoherent claim that there is a perceptual experience of what is not perceptually experienced.

What is this experience of the limits of perception itself? It cannot be perceptual experience. Is it yet another category of experience?

One possible answer is amodality is due to *imaginative* experiences. When we visually imagine a horse, there is in some sense a visual appearance of that horse. However, at least in how it seems, that appearance is nothing like seeing a horse or indeed seeing a picture of a horse.

MacPherson argues that, when it comes to perceptual experience, the content of that experience can come from *imagination*. Imagination plays the role of 'filling in' the gaps that are left out by what is otherwise perceptually experienced. This can explain various phenomena, including some illusions and ambiguous figures. As an example of an ambiguous figure, MacPherson consider the *Necker cube* (Figure 3.1).

The Necker cube (Figure 3.1) resembles a wire-frame cube in which all edges of the cube are visible. The ambiguity concerns how it appears to be orientated in three-dimensional space. It seems to be entirely dependent on the subjective. For example, it seems that one can choose to experience AB as being the bottom line on a face at the front of the cube or at the back of the cube.

MacPherson writes (line AB in my figure matches line AB in her figure),

> [I]if we have two different experiences when looking at pictures of ambiguous figures and this is due to cognitive penetration then my two-step indirect mechanism may explain what happens. Consider the Necker cube [. . .] Perhaps one imagines the line AB either at the front or at the back of the cube, or at least the process by which this imagining would take place is set in motion. Then the imaginative experience or process interacts with one's perceptual process to yield one or other of the experiences of the cube.
>
> (MacPherson 2012a, 56)

The Necker cube seems a particular example of something that appears to be what it is dependent on something beyond what we perceive. If

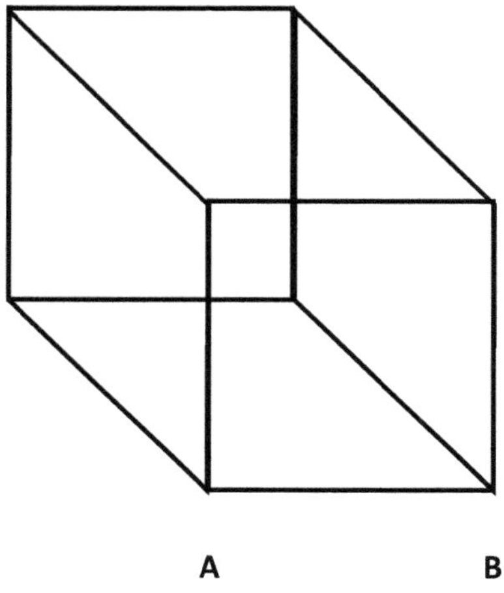

Figure 3.1 Necker cube

we apply MacPherson's analysis to amodality generally, then the kind of experience that is of limits, amodality, and 'presence in absence' is the *imaginative experience*.

However, even if imagination does combine with perception to give us the complete experience that comes with any perceptual experience, there is a problem. We do not seem to imagine the amodal part of perception—the part that goes beyond perception.

In terms of how things appear, what amodality captures is something that is not obviously perceivable. Yet it relates to the phenomenology of perception in that what amodality captures still seems to be *real*. It is part of and completes what we perceptually experience; what we perceptually experience seems to be real. It is an occluded aspect or part, but it is also a real aspect or part of something perceived.

What is apparent in imagination that is obviously imagination is something mind-dependent or *irreal*; it does not appear as if it must be real. In obvious cases of imagining, our experience corresponds to the mind-dependence. For example, I visually or auditory imagine standing on the top of the Eiffel Tower. I am imagining something that is not happening nor has ever happened. In this obvious case of imagining, there is no problem with what I imagine not being how things really are.

However, with the imagination involved in amodality, I cannot so easily assert that what I imagine is unreal. It does not seem to me as if I am experiencing something unreal in what I perceive. It seems to me as if I am experiencing something that is real but also *hidden* or *lying beyond* what I perceive. It does not seem to be something that I see, hear, or touch (which is correct in such cases). Yet it also does not seem to be something that comes from my perception or mind itself.

Thus I resist analysing amodality in terms of imagination due to the phenomenology. This does not mean imagination does not play this role. It only means it plays the role in spite of the phenomenology. This is why I consider the positing of imagination here to be a hidden element of perceptual experience. We are not motivated to posit imagination by the phenomenology.

3.5.3. Representational Structure

Finally, there is the representational structure of perceptual experience. By 'representational structure', I mean the elements, properties, and relations that are necessary and sufficient for perceptual experience to represent something. I assume that we may also call this the structure of the representational *vehicle*. It would be a mistake to call it the representational *content* of perceptual experience. The difference between the vehicle and content is why I consider any representational aspect of perceptual experience to be hidden (and intentional).

I assume the following about the difference between the vehicle and content of representation:

- *The vehicle represents the content.* The vehicle must be real for the perceptual experience to be a representation. If there is no representational vehicle in an experience, then an experience is not a representation.
- *The content is represented by the vehicle.* It is what the perceptual experience is *about*. The content must be represented in some way. If no content is represented, then there is no representation going on. However, the content need not be real for the perceptual experience to be a representation.
- *In most cases, the content is identified with the perceptual appearances or phenomenology.* However, it is conceivable that more is represented in experience than the phenomenology. Those who resist identifying representationalism content with phenomenology argue that there is more to phenomenology than representation.
- Finally, the vehicle can resemble but need not resemble the content. Assuming a resemblance, especially via identity, is to confuse vehicle and content. As such, given the content, the vehicle might be structurally identical to the content—e.g., with the same temporal properties or colour properties—but it need not be.

I make these assumptions based on various claims by perceptual theorists known as *representationalists*, those who hold that the features of perception, including perceptual experience, can be explained by the representational structure (e.g., Lycan 1996, 2008). For example, of attempts to link Husserl's phenomenology of time consciousness and neuroscience, Grush writes,

> In all cases, the core of the proposal is the same—(i) this or that theory or analysis from some area of cognitive science or neuroscience imputes a certain kind of structural feature to patterns of neural activity; (ii) language that can be used to describe these structural features can [. . .] also be used to describe aspects of the phenomenology of time consciousness as Husserl analyzes it; therefore (iii) some explanatory or otherwise interesting connection between neural mechanisms and phenomenology has been revealed. *And a diagnosis in all three cases is the same: content/vehicle confusion. The material in (i) concerns vehicle properties, never content properties, while the material in (ii) concerns contents, never vehicles, and so one can get a direct route from (i) and (ii) to (iii) only via confusing vehicle and content.*
>
> (Grush 2006, 418;[13] emphasis mine)

What matters most here about representationalism is what conditions it sets on the structure of perceptual experience, especially given the phenomenology of perceptual experience. As far as I can tell, it sets no conditions on the structure.

One may confuse the vehicle and content in one's theory. When it comes to the structure of experience, the constraints on that structure from taking content seriously are obvious. These constraints are those that come from appearances, from the phenomenology. I already discussed them when discussing naïve realism. However, constraints from a representation considered independent of the phenomenology are not at all obvious. For example, constraints due to the nature of the vehicle are not at all obvious.

As far as I can tell, conceiving of experience as representational offers nothing to questions about the structure of experience intrinsically, relationally, in space, or in time. Any variation of such structure is compatible with experience being a representation.

Representation does not help to distinguish the difference between kinds of errors in perceptual experience given different views of time. Since it does not constrain how perceptual experience is embedded or extended in time, it does not constrain how perceptual experience fails to be constituted by what appears to be perceived.

It might constrain the structure in some way if the representational vehicle itself were part of the phenomenology. Then, like any other phenomenological features, it would involve some commitment to the appearances in order to have veridical experience.

However, most representationalists hold that the representational structure is a non-apparent feature of perceptual experience. The representational aspect of experience is *transparent* or *diaphanous* (Crane 2011); the representational elements are not apparent; they are not anything that are experienced themselves; they are hidden features of experience. Logue writes,

> It's important to keep in mind that [the representationalist] doesn't hold that a subject perceives the representation in which her experience fundamentally consists [. . .]. [T]he representation [. . .] is a state of the subject that enables perception of things in one's environment without being perceived itself, much as the eyes see without being seen by its subject.
>
> (Logue, *op.cit*, footnote, 208)[14]

This hiddenness may even be necessary for representationalism about experience. Alston writes that one reason for representation about perceptual experience might be that perceptual experience "is experienced as a representation." However,

> that is clearly not the case. When something I see looks a certain way to me (conical, red . . .) it doesn't appear on the face of it be a representation of anything. [. . .] Phenomenologically it has the character

> of a presentation of an object as being such-and-such. The experience terminates in the object presented without, so far as it appears, functioning to put S in mind of something else. Hence we need a reason beyond the phenomenological character of the experience to take it to be a representation.
>
> (Alston 2005, 276)

We might still have it that the representation constrains the structure. Again, if the structure of the vehicle is the structure of the content, then, even if the representation is not part of the phenomenology, we can still constrain the structure based on the representation. Because of its relationship to the phenomenology, we can take the structure of the representational vehicle from the appearances and use the vehicle's reality to make inferences about the actual world.

However, as stated, there need be no resemblance between representational vehicle and content. As such, we cannot infer the structure of the representational vehicle from the phenomenological structure. They need not be the same.

That is, one cannot infer the real representational structure of an experience through the content until, at least, one knows how they are related. We do not get this relationship simply by characterising experiences as representational. We need more.

I think we need an account of the structure of experience in space and time, and of how such structure relates to the structure of what is experienced (what is perceptually apparent or content). This affects our interpretation of experience as being erroneous. For example, the temporal structure of experience determines when the experience happens; if we know that, and know when the content happens, we have some idea of how they are related. Given perceptual appearances, these should be the same. If there is a difference, then our perceptual experience is erroneous. An account of the nature of this difference can help us understand how this experience is erroneous.

To give this account, one should pay attention to the empirical evidence of error. However, one should also pay attention to what is true about the nature of time and how experiences and their content can exist in it. That is, to get a true relationship between representational vehicles and content, one should look at experience from the true philosophical theories of time.

3.6. Conclusion

We have now discussed perceptual experience under several headings:

- The varieties of experience, of which perceptual experience is one.
- The structure of experience, as motivated by naïve realism and some considerations about space and time given tenseless theory about time.

- The relations involved in the structure of experience, again motivated by naïve realism.
- Possible hidden structure of experience (structure not motivated by naïve realism). This includes subjective activity, imagination in amodal perception, and representational vehicles.

In this chapter, I have usually assumed veridical experience. I have also already brought in some analysis that is motivated by the philosophy of time. However, to assume there is only veridical perceptual experience is far more naïve than even naïve realism. There is at least the possibility of erroneous perceptual experience.

So how possible is it that our perceptual experience is erroneous? Does time matter in working that out? To see if time does matter, we must first engage with what kinds of erroneous perceptual experiences there could be. This is the subject of the next chapter.

Notes

1. I neglect some things that others may consider important. Notably, I ignore considerations from a characterisation of experience as representational. At the end of this chapter, I briefly explain why I do this.
2. A similar case that might help clarify this point does not concern imaginative experience but concerns actual images in the external world, such as pictures or films. Images have some kind of representational or intentional aspect to them (e.g., Sartre 1986; Blumson 2009). However, they can resemble things that do not have such an aspect. First, a non-imagistic x can resemble an image of some non-x without being such an image—e.g., a cloud can resemble a sculpture of an animal yet be neither a sculpture nor the represented animal. Second, a non-imagistic x can resemble an image of itself (that is, of x) without being such an image—e.g., I might think I am looking at a picture of a person and it turns out I am looking through a window.

   If I see what appears to be such an image—e.g., it appears to be a picture—and it is just as it appears to be except that it is not an image, then I might have an erroneous belief about what I see. In that case, perhaps, I also have an illusory experience. However, the only experience that seems relevant here is my seeing of the thing I am mistaken about; in the example, how it visually appears is not mistaken. As such, I am unsure if I would call this an illusion.
3. An objection might be that perception is not a perception of the perception itself. Perceptions are not reflexive or self-reference. Thinking that is just a mistake.

   The account here is not such a mistake. It does not involve proper reflexivity or self-reference. The perception is not of the *whole* experience. What we perceive, perceived objects or perceived events, partially constitute the perception. We perceive part of the perception, the part that is apparent to us.
4. Some theorists take this to mean that perceptual experience is *transparent* or *diaphanous* (e.g., for discussion, see Pace 2007; Molyneaux 2009; Crane 2011). It is not apparent to us. Only the content—what we perceive—is apparent to us.

   I take it that this conception of perceptual experience is as only something subjective that happens to us. However, given naïve realism and considerations about experience's constitution, I do not conceive of it as only being subjective; it also includes what we perceive (and that is not transparent).

5. Naïve realism does not work for other varieties of experience. For example,

    - We imagine unreal things. If what we imaginatively experience is what we imagine, then we experience something unreal. If this is how things appear, then things seem unreal. This is not naïve realism.
    - We remember past things. To be naïve realist about this, we would need to deny certain theories of time and do based on the phenomenology. If naïve realism about memory were true, then, for there to be a memory-experience of past things, past things would need to be real. That is, eternalism would need to be true and presentism false. As such, that there is memory-experience coupled with an account of experience's constitution settles a debate about time. Presentists are likely to consider this too fast. Anyway, it is not obvious that remembered things are components like this (I discuss this later).

    For such reasons, we may hold that some experiences do not include what we experience as a component. As such, naïve realism does not apply to them.

    However, it is worth noting why naïve realism does not apply. It is not because it is naïve; it is because it is *realist*. We are trusting experience; we are not denying. Similarly, trusting perceptual experience, if naïve realism is right, what we experience is part of the experience.
6. For example, Chalmers and Clark (1998) and Rowlands (2010).
7. The position that it is only composed of entities internal to the subject and to the subject's actions seems more like *enactivism* about perceptual experience. Indeed, according to Jacob (in Matthen 2015), enactivists "endorse a version of what has been variously called 'the extended mind thesis', 'vehicle externalism', or 'active externalism' " (Jacobs (in Matthen 2015, 228)).
8. In addition, as far as I know, there is no spatial debate analogous to any of these temporal debates. For example, there is no serious philosophical position about space that is analogous to presentism—i.e., a position that holds only one point in space is real.
9. In my 2010, I call these elements *phenomenal parts*. As may be noticed, I refer to them as parts, still, in this book. However, I no longer qualify them as phenomenal because I have not found a widely settled use of that term in contemporary philosophical thought.
10. Note that there is a causal theory of perception going back as far as Locke and more recently and most influentially developed by Grice 1961. I don't consider this here because it does not seem clear to me that it either supports or denies naïve realism or the kind of constitutional account of experience I examine here. This does not mean it could not be adapted to this discussion, but I do not adapt it here (see also Coates 2000; Arstila & Pihlainen 2009).
11. Robinson argues that intentionality is irrelevant to perception (Robinson 1974).
12. Some theorists also make a distinction between intentional content and intentional objects (for discussion, e.g., see Crane 2001). For the purposes of this work, I refer to both as intentional content.
13. Later in the paper, Grush advances some indirect routes. I discuss such approaches in the chapter on temporal experience.
14. In the expanded note, Logue refers to this position as *intentionalism* and distinguishes it from another representational view, sense-data theory. Sense-data theory is the view that we experience intermediary sense data, and those are

apparent to us in some way. Thus this kind of representationalism would be a case where the representations are experienced. However, few representationalists these days seem to hold sense-data theory. Furthermore, still in keeping with Alston's point, even if we do experience representations such as sense data, they do not appear to be representations, but just parts or aspects of things in the world.

4 Erroneous Experience

> It's not uncommon for someone waking from a coma to hallucinate. But how to describe "hallucinate" to the Crakers? It's when you see something that isn't there. But if it isn't there, Oh Toby, how can you see it?
>
> (Margaret Atwood, *MaddAddam*)

A popular topic in my classes is illusion. Although certain conclusions and arguments in other sections of my courses may surprise them, they do not have the immediate 'wow' factor elicited by examples of mistaken perception. Once I point out the error, 'perceptual illusions' of movement, colour, shape, depth, contrast, and so on are immediately interesting to those who suffer them.

Many of the examples of mistaken perceptions, especially 'perceptual illusions', are visual—e.g., the *Müller-Lyer*, *Hering*, and *Ebbinghaus* illusions (Figure 4.1).

These are perceptual experiences of simple geometric figures, experiences of which are in some way mistaken. As these phenomena involve geometric properties, I follow Wackermann, and refer to them as 'Geometric-Optical Illusions' or *GOIs* (Wackermann 2012).[1]

With respect to the specific examples, the mistake in the GOIs are as follows:

- *Müller-Lyer*: The central lines in the figure appear to be different lengths; they are not different lengths.
- *Hering*: The vertical lines appear to bend away from one another; they are parallel.
- *Ebbinghaus*: The central circles appear to be different sizes; they are the same size.

Other phenomena commonly classed as illusions are the tricks of professional magicians and illusionists. A magician stands on a stage in front of a black background under very bright light. She holds up a bright flower

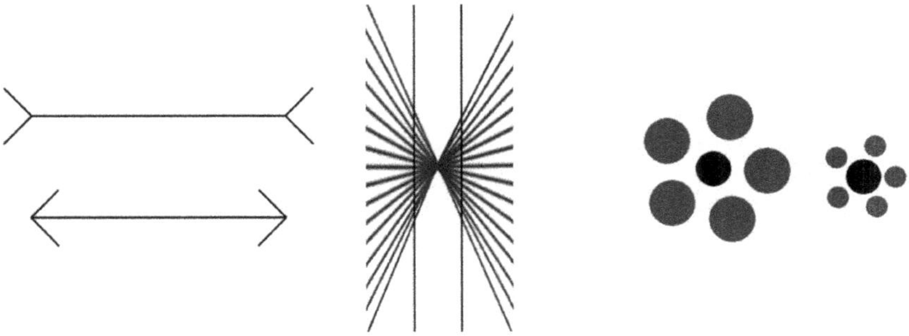

Figure 4.1 Müller-Lyer Illusion, Herring Illusion, and Ebbinghaus illusion

and waves her hand. Once she does so, the bright flower leaves her hand and floats across the stage. The illusion is that it seems to do so under its own power; there are no obvious strings or any attachments to the magician. So far as one can tell on looking at it under this bright light, in front of this dark background, this inanimate flower has the power of self-propelled motion (this example comes from MacKnik & Martinez-Conde 2010).

I refer to these examples as being *called* 'illusions' rather than simply *being* illusions (without quotes) because, despite how startling these phenomena might be, there may be reason to hold that these are not illusions at all, at least not in a philosophical sense. Even if they are illusions in a philosophical sense, there may be reason to hold that they are not illusions of the properties that they are commonly thought to be illusions.

The classification of something as illusory (or otherwise erroneous) in part depends on phenomenology and in part on metaphysics. In the coming chapters, I argue that the case for a particular instance of experience to be illusory (or, generally, an erroneous experience) depends (a) on how such an experience extends and is embedded in time and (b) on how the constitution of experience allows or disallows naïve realism.

However, before doing this, the concepts of illusions and other kinds of erroneous experiences need some clarification. This chapter argues for some conditions for illusion, hallucination, and anosognosia, and distinguishes between them.

First, I outline the conditions of illusion. Second, I distinguish illusion from other kinds of erroneous experience, in particular, hallucination. In addition, I expand the taxonomy of erroneous experience beyond the distinction between illusion and hallucination. I make further distinctions in illusion itself, bringing in complex illusions and sortal illusions. I also introduce and develop anosognosia, and a specific kind of it: conflationary error.

4.1. What Is Common to Illusion and Hallucination

Illusions matter to naïve realism. A common claim is that an illusion of x is a counter-example to naïve realism about x. Recall that naïve realism about perception holds that perceptual appearances correspond to reality—e.g., if there visually appears to be a green cube over there, then, if naïve realism is true about green cubes over there, there is a green cube over there. However, if there is an illusion of a green cube over there, there is no green cube over there; naïve realism is false for green cubes over there.

This is an example of what is often called the argument of illusion against naïve realism (e.g., Furlong 1954; Dancy 1995; Reynolds 2000); one must deny naïve realism because there are illusions. A reverse argument may be thought also possible: One must deny there are illusions because naïve realism is true.

However, the latter argumentative strategy seems an almost impossible task—at least with respect to a denial of there being any illusions at all. No matter one's ontological or metaphysical commitments, some appearances must fail to match reality. For example, at least some erroneous experiences are of mind-independent spatial depth (such as the apparent location of purple spots on a ceiling). At least some erroneous experiences are culture specific (e.g., whatever Saint Anne looks like, she is unlikely to look like her renaissance portraits).

Perceptual erroneous experiences are frequently categorised under one of two kinds: illusions and hallucinations. In contemporary philosophy of perception, illusion and hallucination are almost without exception treated as if they are distinct kinds of error. Different kinds of evidence are advanced in support of one or the other and different theories of perception are evaluated in light of each kind of error. Some philosophers weigh perceptual theories against each other based on this distinction. For example, Smith argues that a naïve realist variant of disjunctivism may explain hallucinations, but it cannot explain illusions. In contrast (he goes on), a representational variant of disjunctivism can account for both. Thus he recommends disjunctivists adopt the representational variant (Smith 2010).

This suggests that a theory of perception should include an explanation of both illusions and hallucinations, and that there should be a categorically different way to explain each of them. However, there are two questions about such a distinction between illusion and hallucination when it comes to explaining actual experience.

(1) Are the conditions for illusion and hallucination mutually exclusive? Erroneous experiences fall in one, or the other, but not both. When we know an experience satisfies one of these sets of conditions, then we can place it in that category and not the other. If something is an illusion, then it is not a hallucination, and vice versa.

(2) Is any actual (not merely possible) experience a member of either category of experience? We might have a clear distinction between illusion and hallucination (answering the first question), yet there may be no examples of illusion or hallucination. Each of these may be kinds empty of actual instances of experience. (It is plausible to hold that some categories of experience are actually empty—e.g., naturalists should hold divine experience to be a category of experience that has no actual cases.)

Number (2) is important. Unlike veridical experience, there is no obvious epistemological or folk reason to demand that there be hallucinations or illusions. It is not a failure of a coherent theory that it can give an account of actual experiences, including their appearances and their phenomenology, but without appeal to either illusions or hallucinations. (From an empirically minded perspective, I am tempted to say that this is a good feature of a theory.)

If we answer 'no' to either of these questions, then we do not have reason to require theories of perception to explain illusion and hallucinations separately. Either these are not distinct categories or no actual experience falls into them.

Before deciding the differences, I consider some common features they share.

4.1.1. Appearance Is Different to Reality

The first obvious feature common to illusion and hallucination is this: For both illusions and hallucinations, appearance is different to reality. This is a necessary condition of both: If we want to show there is either an illusion or hallucination here, we need to first show that there is a difference between appearances and reality. If we cannot show this, then we cannot show that our experience is either an illusion or a hallucination.

However, although a necessary condition of both illusion and hallucination, it is not a sufficient condition for either. There are cases in which a difference between appearances and reality are due to there being more to reality than appearances. This is anosognosia, which I discuss in a later section. In that case, although there is still error, most of what is apparent is also real; however, there is more to reality than what corresponds to the appearances. The error is that there does not appear to be more. With respect to the particular thing experienced, there seems to be nothing more to reality than what is apparent.

Perhaps the sufficient difference is this: What is apparent is not real. For example, it may be that I seem to see aliens. However, there are no aliens—aliens are not real—but there seem to be aliens. Thus I am under an illusion or hallucination of aliens.

For perceptual experience, this is sufficient. With perceptual experience, if it seems that there is an X, but X is not real, then there is this discrepancy between how things seem and how things are. However, this is also not a necessary condition for perceptual experience. What we seem to experience that, in fact, is illusory or hallucinatory, can be real. I may seem to see aliens, and there may even be such aliens, yet I cannot seen them. This is because the aliens are too far away for me to see. If they are so far away, yet it seems to me that I see them, then this is an erroneous perception. However, the aliens are real (just far away) and they are apparent (I sure seem to see them).

Generally, in terms of structure, we can say why it might not be possible to see something that we seem to see and is real, and thus to have erroneous perceptual experiences of it. It is because the structure of experience itself prevents us doing so. How the experience is constituted, embedded, and extended in space and time prevents our perception being of such things. Thus, even if they are real and apparent, we are still under an illusion or hallucination of them.

For example, one that immediately brings time into consideration, say aliens evolved only within the last million years, and the planet they are on is one million and one light years away. Assume that my perceptual capacities do not involve faster-than-light relationships between what I see and myself. My perceptual experience cannot be constituted by things separated from its subjective activity by things that are at distances light is too slow to cross.[2] I cannot perceive them because they are too far away.

In demonstrating that some experience is an illusion or hallucination, I need to show that I cannot experience the relevant entity. This turns on giving an account of the structure of perceptual experience.

Thinking of perceptual experience in terms of both its content and constitution, and in naïve realist terms, is helpful to understanding why it might or not be an illusion or hallucination. For some content X, we can ask if it fails to match appearance and reality. If it fails, x is not a constituent of the experience and thus is not experienced.

4.1.2. *Illusions/Hallucinations Need Not Surprise nor Be Believed*

That an experience is surprising or rare does not make it an illusion, a hallucination, or any other kind of erroneous experience. I assume that a necessary condition of a surprising experience is that one does not expect it to occur. I assumed that one's expectations about experience are based on one's background beliefs about the possibility of experience. As one's background beliefs can be wrong, one's background beliefs about the possibility of experience need not correspond to the possibility of experience.

Similarly, rare experiences are not erroneous merely because they are rare. Neil Armstrong's visual experience of the moon's surface is not an erroneous experience of the moon. Few people have experienced the summer solstice at the North Pole.

Nor need an illusion or hallucination be a surprising experience. For example, anyone working on perception can become accustomed to perceptual illusions. That does not make their experience any more accurate. This leads to a general point: In order for us to be under an erroneous experience, our beliefs need neither correspond nor be in conflict with appearances. It can seem to us as if something is perceptually apparent, it is not actual, and we do not believe it to be actual. For example, take any of the illusions at the start of this chapter.

However, because of our beliefs, we may interpret experiences as erroneous or as veridical. As such, because of our background beliefs, we may mistakenly believe an erroneous experience is veridical or a veridical experience is erroneous.

That is enough on what illusions and hallucinations have in common. Let us turn to each of them individually in more depth.

4.2. Illusion

What is an illusion? One understanding of illusion is a kind of false belief. One could align it with false beliefs generally, or particularly pervasive, or even necessary ones (e.g., Nietzsche's 'Apollonian Illusion'). However, in the philosophy of perception, illusions have a much more restricted understanding. Roughly, they are cases where we seem to experience properties (or relations) that we cannot experience. This can be because there are no such properties but it may also only be because we are unable to do so due to the structure of experience itself.

This may seem to be enough for illusions. If something perceptually appears to have a property but does not have that property, then there is an illusion. However, there are a number of issues with this.

We might ask what stands as a property here. Is it anything at all that fits a descriptive predicate? Say there is a small statue of a cow that, because of where it is placed relative to an observer, appears to be a life-size faraway statue of a cow. Say we look into a painted room which contains only a number of perfectly camouflaged chameleons and so appears to be (but is not) an empty room?

Are these illusions of *being faraway* or *being empty*?

With respect to *being faraway*, I accept that there is an illusion. I accept that there can be illusions of relations, and this may be an appearance of a relation of distance (say, of a large distance) between the statue and the experiencing subject.

The case of *being empty* is more complicated. We might call emptiness a *negative* property. It is a property defined by not being something else; that is, it is the property of *not being filled (or not being occupied* or similar). Any property not defined this way is a positive property—e.g., redness is a positive property; it is not merely *not something else*; it is not merely *not green.*[3]

Negative properties raise two questions:

(1) Do we have experiences in which there are erroneous appearances of such negative properties?
(2) Is there a better way to explain errors of such negative properties in experience other than illusions of their appearance?

Later, I argue that we can answer question (2) alone with anosognosia. However, to give that account, I must answer question (1) in the negative: We do not have erroneous appearances of negative properties. I do not have a satisfactory argument for that answer. I only have the intuition (for all that is worth) that there are no erroneous appearances of negative properties—not because they are not *erroneous* but because they are not *apparent*. My intuition is that all there is to the appearance of negative properties is the non-appearance of another property—e.g., the appearance of an empty space is only a *non-appearance* of filled space.

With that aside, I take this position: If there is an appearance of a property (that is not a negative property), without there being an instance of that property, then we have an illusion of that property.

This may seem to be the same as other philosophers' definitions of illusion. For example, Kalderon defines illusions as follows:

> [A]n illusion is an experience of an object o appearing F where o is not in fact F.
>
> (Kalderon 2011, 4)

For example, if we substitute the colour green for 'F' and some object in the world—e.g., a cube, for 'O'—we get the claim that an illusion is an experience of a cube appearing to be green where the cube is not in fact green.

My definition of illusion is different to Kalderon's definition. I do not find Kalderon's definition sufficiently precise for my purposes. Although his definition marks a discrepancy between appearance and reality (the 'in fact') with respect to a property, it does not clearly mark what the discrepancy might be. Given that there is an object, one might think that the discrepancy has to be in the property; the appearance and reality of the property are different—e.g., there seems to be an instance of the colour green and there is no instance of the colour green. However, this is not the only way to unpack the discrepancy between appearance and reality.

Another way there might be a discrepancy is due to a class of erroneous experience I introduce here: *conflationary errors*.

4.2.1. *Conflationary Errors*

Here are two ways to understand Kalderon's definition of 'O appearing F' yet O not being F. Both involve error but only one is an illusion of F. Call these X-*errors* and Y-*errors*:

X-error

(i) O appears to be F.
(ii) There is no F.
(iii) O is not F.

Y-error

(i) O appears to be F.
(ii) There is an F.
(iii) O is not F.

In both cases, one might think F is illusory. Given an X-error, this is correct; there seems to be F, but there is no F. However, given a Y-error, there *is* F. We might very well experience F. It is just that the following is not the case: F is not a property of O. It only appears to be a property of O. That is a mistaken appearance. However, this (a Y-error) is not necessarily a mistaken appearance of F.

It may be that another object Q is F. It may be that we perceptually experience Q (and Q or some property of Q is a constituent of our perceptual experience). This is why we have the appearance of F. In that case, we do veridically perceive F. For F, we ought not to say this is an illusion of F. Instead, there are two other possibilities.

First, there only *seems* to be O. There is no O, or, even if there is an O, we cannot perceive it. In that case, what is illusory is the appearance of O. Thus if this is any kind of illusion, then it is an illusion of O (or, as I discuss later in this chapter, it is a hallucination of O).

Second, as well as F, we also perceive O. However, O is not F. It just appears that O is F. In fact, Q is F, and we perceive that instance of F that Q possesses. In addition, we perceive O. The mistake is that we perceive O and F together in some way such that O seems to be the possessor of F. This is the mistake; this possession is the discrepancy between appearance and reality.

However, this mistake is neither a discrepancy between the appearance and reality of O nor a discrepancy between the appearance and reality of F. Again, there is O and there is F; we are perceptually experiencing O, and we are perceptually experiencing F. Instead, the discrepancy concerns what has the property.

Say that I am holding a ball in my hand. I feel and see a ball in my hand. The ball feels quite light; I can even toss it lightly in the air. As I make this action, I also see a ball in my hand bouncing about. Then I see it fall off my hand (and feel it disappear from my hand). It drops on my foot, almost crushing it with a hard heavy weight.

I am not under an illusion of the weight of the ball that I feel myself throwing into the air or the ball that I see fall onto my foot. I am under an error of *conflation*. The ball I feel on my hand and the ball I see on my hand are two entirely different balls. One I feel in my hand and feel leave my hand as I throw, another I see and later feel crush my foot.

What is mistaken is my experience of a *single* ball that I believe I feel and see, feel and see it thrown out of my hand and falling heavily on my foot; this one ball does not exist. There is no single ball with all of these properties.

Let O be the object that is the ball I *see* in my hand; let its rolling off my hand be the property F. Let Q be the object that is the ball I feel in my hand; let its weight in my hand be the property G. Let P be the object that is the ball I seem to see and feel; P is a lightweight ball seen rolling off my hand that then causes a crushing pain in my foot.

Thus it seems that I see a single object P that is F and G. I do not see P that is F and G. That is the discrepancy between appearances and reality. However, I do see some object that is F and some object that is G. It is only that these two objects (O and Q) are not the same object (P). I am conflating two objects into one:

- One ball (Q) is a heavy ball that rolls from (what seems to be) my hand. It is not the ball I feel in my hand (that is O). However, I do feel Q a moment later on my foot.
- The other ball (O) is a lightweight ball that I feel in my hand. However, I do not see O roll off my hand or feel O land on my foot (that is Q).

As a result, I have a Y-error kind of experience. I misidentify two objects (O and Q) separately possessing two properties (O is F and Q is G) as the same thing (P) possessing these properties (P is F and G).

It is not that some property appears which is not real, or even that a multiple of properties appear which are not real. It is that I identify them as belonging to the same thing. How the properties are collected together (are possessed, borne, or bundled) is what is mistaken. I mistakenly *conflate* a multiple of objects into one.

I call this error of an experience a *conflationary error.*

As a final note, conflationary errors need not happen with respect to appearances of single things. Multiple objects might seem to bear properties that do not, in fact, possess them (yet something does).[4] In order that such conflationary errors are not cases of illusion, they must satisfy a number of conditions.

Let the set of properties that we appear to perceive be *AP*. Then the conditions for conflationary errors are as follows:

- We do perceive *AP*. Thus these experiences are not illusions of *AP*.
- The objects that appear to possess *AP* do not possess *AP*.

In that case, the questions move from questions about how an illusion of some properties is possible to different questions about conflationary errors:

(a) What really has the apparent properties?
(b) Why do there seem to be these objects with these properties?

I consider conflationary errors a kind of anosognosia. The important point here is that there need be no illusion of F here; there need not even be an erroneous experience of O. There is only the mistaken appearance of *O being F*.

4.3. Hallucination

We have briefly raised one possibility for why there is a discrepancy between the appearance of O being F and the reality that O is not F. This is where, although there is F, there is no O. So is this appearance of O, which is not the reality of O, an illusion of O? This depends on what we mean by O.

As with illusion, there seems to be more than one possibility:

(a) O is the set of properties, or relations, or the kind belonging to the object that seems to bear F.
   For example, if a ball seems to be green and light, is light, but is not green, then we have this: F is the greenness; as we are talking about a ball, it is spherical (let us say that it has 'sphericity') and is light. Thus O is the sphericity and lightness of the ball, or the relations between its parts, or the kind of thing it is—i.e., the ball.
(b) O is the object *itself* that bears F.

If an illusion is concerned with the appearance of properties that are not real, only an erroneous experience of one of these might be called an illusion, and that is (a). It is an illusion of properties and relations. In this model, it concerns the appearance, but not the reality of shape, weight, etc.

Yet, as with the example earlier concerning illusions, this discrepancy might even not be an illusion of these properties. There are such properties, and we do experience them. Instead, we conflate the bearers of these properties with the bearers of F (as for whether or not this is a case of illusion, this is a question to which I return later).

Option '(b)' is an erroneous experience, but it is not of properties, relations, or kinds. Option '(b)' is of the object that has the properties, relations, and kinds. If we only have an appearance of the object, but there is no object—i.e., as in '(b)'—then this is a hallucination.

Typically, a hallucination in philosophy is any experience in which (a) there is an appearance of an object but (b) there is, in fact, no object. This distinguishes it from illusion—where it is a case of property, not object.

However, I must make such a distinction more precise. I need to work out the following: What distinguishes the appearance and non-occurrence of objects from the appearance and non-occurrence of properties? What is the *object* that is the object of hallucination?

4.3.1. The Object of Hallucination

Johnston (2004) calls the object of hallucination the 'obscure' object of hallucination, suggesting that what such an object 'is' is difficult to identify. Hallucinations are not experiences of things in the world. An hallucinated thing is not something perceived, even in a distorted way (as in mirages or illusions). It is not some kind of special mode of experience or experience of special things—e.g., as religious or mystical experience might be a special mode of experience (e.g., we have a mystical experience of a common tree in a park, perhaps).

Philosophically, the common understanding of hallucination is of an experience of some object that appears to be there but is not there. This condition is commonly used to differentiate hallucinations from other kinds of perception, both veridical and erroneous—e.g., from illusion. For example,

> we sometimes seem to see objects as other than they are (illusion) or seem to see objects that are not there (hallucination).
>
> (Fish 2009, 25)

> Perceptual experiences are often divided into the following three broad categories: veridical perceptions, illusions, and hallucinations. For example, when one has a visual experience as of a red object, it may be that one is really seeing an object and its red colour (veridical perception), that one is seeing a green object (illusion), or that one is not seeing an object at all (hallucination).
>
> (Soteriou 2010, section 1)

There is another way to describe hallucinations, one that does not engage with the reality of what we perceive. Instead, it only insists that the hallucinating subject does not perceive the thing. For example, Crane writes,

> A hallucination [. . .] is an experience which seems exactly like a perception of a real, mind-independent object, but where there is no mind-independent object of the relevant kind being perceived.
>
> (Crane 2011, section 1.3)

Central to both definitions is that the subject seems to perceive an object which they do not perceive, either because it is not there (whether or not they could perceive it) or because the subject cannot perceive it (whether or not it is there).

Say I look at my arm and see a long string of what seems to be DNA—i.e., a helical structure. Assume that I cannot see the complex structure of my own DNA by looking at my arm. Although there is, in fact, something there (my arm), and even something resembling what I seem to see (many things: the multitude of DNA strands constituting my arm), yet unless I do in fact have powerful perceptual abilities, I do not see this strand. I do not in fact have powerful perceptual abilities. Thus the DNA string I seem to see is not what I see. I visually hallucinate it.

4.3.2. Hallucinated Things versus Internal Things

What might complicate this definition is that, given indirect realist theories, we do perceive something. We perceive an *internal* entity. Indirect realists claim that we do so even in the absence of sensory stimulus, even in cases where we conclude we are suffering hallucinations. Indeed, the cases of hallucinations are one motivation for indirect realism. However, this means that, strictly speaking, we do perceive something. We perceive a sense datum or a set of sense data; we might even say that sense data are real. It is only that we do not perceive *external* things.

However, this is not much of a problem. Whatever internal things we do perceive might be, they are not the hallucinated things that we seem to perceive. For example, we may see colour and shape sense data; these may be real; these may even be realised in our brain (or not); their presence may even be why we hallucinate. However, if seeing these sense data means we hallucinate something external, then it is the case that we do not see the external thing. For example, we see an internal green sense datum and round sense datum, and this means we hallucinate a green ball outside of us. We still do not see the green ball.

Central to the concept of a hallucination is that no object that we hallucinate is out there, or, if it is out there, we cannot perceive it, yet there seems to be some object out there, indeed perceptibly so. The appearance yet imperceptibility of an object marks the difference between illusion and hallucination.

Thus we can generally state the difference between illusion and hallucination as follows:

- *Hallucination*: It appears as if there is some object (O). There is no O or we cannot perceive O. This is a hallucination (of O).
- *Illusion*: Some object (O) appears to have property F. There is no F or we cannot experience F. This is an illusion of F.

4.3.3. Hallucinations Are Not Mirages

Central to the philosophical concept of hallucination is that the hallucinated object appears somewhere yet either (a) there is object there and/or (b) whatever object is there is imperceptible to us.

This definition of hallucination turns on the idea of an object being there or not being there. However, this raises the possibility of confusion with another kind of erroneous experience: mirages.

Hallucinations are not mirages. Hallucinated objects are not simply objects that seem to be in one place but are actually in another place. This might be a case of mirage or a case of illusory spatial location (if there is a difference (as discussed later)). With hallucinations, the hallucinated object is either nowhere or, if they are anywhere, they are nowhere perceptible.

The spatial properties of an object are a property: that a cube is at this distance from me, or occupying this volume, is to describe the cube as having certain spatial properties: distance, location, and size. These may be mistaken. These may be mistaken because of the paths light takes to get to us—i.e., what gives rise to mirages. In addition, it may be due to something within us (and so require explanation in terms of illusion). However, so long as there is the cube, and we perceive it, there is no hallucination here.

In saying that there is nothing there, we can have two understandings of this, one of which is a mirage or illusion of spatial location, the other a hallucination.

In order to be a mirage or illusion of spatial location, it can be that there is an object (O) and a spatial location (T), such that

- O appears to be at T.
- 'There' is the location at T. When we say O appears to be there, we mean it appears to be at T.
- *We perceive* O.
- However, O is not at T. O is not there.
- Therefore, we have a mirage or illusion of O being T.

In order to be a hallucination, it must be that there is an object (O) and a spatial location (T) such that

- O appears to be at T.
- 'There' is the location at T. When we say O appears to be there, we mean it appears to be at T.
- *We do not perceive O (either it does not exist or we cannot perceive it).*
- Therefore, we have a hallucination O at T.

The condition for hallucination is not that O is not at a particular location. It need not be there. Either there is no O anywhere at all or O is not perceptible.

Yet there is another condition here that must be met: There appears to be an object O. This raises the next issue. What does it mean to say that an object is apparent? Answering this is less simple than it initially appears.

4.3.4. Apparent Objects and Property Theory

The general question 'what does it mean to say that an object is apparent' seems easy to answer (if any question about appearances can be easily answered). Just look at anything and see. Look at an apple, or your own hand. Is it apparent to you? Well, it is an object. Thus if you know what it is for this apple or your hand to be apparent to you, then you know what it is for an object, or at least one object, to be apparent to you.

However, do we see an object we call an apple or a collection of properties we call an apple? That the answer is one or the other depends on there being an apparent distinction between the object and the sum of its properties.

That there is a distinction at all is a matter of metaphysical contention between substance theorists and bundle theorists. Substance theorists hold that concrete particular things are a collection of instances of properties (and/or relations) borne and bonded together by an underlying substance, substrate, or bare particular. Bundle theorists hold that they are only a collection (or 'bundle') of instances of properties which happen to share spatiotemporal location (or are otherwise compresent).

This debate is a metaphysical debate. There is no phenomenological, no empirical, no apparent difference if one side or the other side is correct. If that is right, then there is no apparent difference in an object and a collection of instances of properties. As such, one cannot distinguish the appearance of collections of properties and the appearance of an object.

If one cannot distinguish them, then there is an alternative interpretation of what one might be tempted to call hallucinations. It is this: We are under illusions of instances of multiple properties, and nothing more (albeit, that is a lot in itself). I call illusions of multiple properties *complex illusions*.

4.4. Complex Illusion

Let us define complex illusions as follows: They are cases of illusion where a multiple of properties are illusory.

For example, it may be the case that what I perceive seems to be a Rubik's cube: To be cubic, to have multiple colours, to have independently manipulable parts, and to be hard to the touch. However, what I perceive is really spherical, one colour, no manipulable parts, and soft. If I only perceive the one object, I am misperceiving its properties. I do not perceive its actual properties. I am under an illusion proper of the properties it seems to have. In this example, I perceive what seems to be a Rubik's cube but, in fact, I perceive a rubber ball.

As with illusions generally, I need not, in suffering a complex illusion, perceive a multiple of properties belonging to different objects as properties belonging to one object. For example, I need not see, e.g., a single-coloured hard cube, a multi-coloured hard sphere, an amorphous colourless soft blob, mistakenly appearing to be one thing. It need not involve conflationary error.

However, by this definition, I am only misperceiving properties. I am not misperceiving the object itself. Again, I perceive the object but as having properties it does not have.

I might perceptually experience (i) the object and (ii) its properties. I do not perceive 'ii'. Yet what is left to correspond to appearances once I eliminate 'ii'? What is the appearance of an object that is not just the appearance of its properties?

This is important for the distinction between hallucination and illusion. If we hold that (i) a hallucination is not just a kind of illusion and (ii) there seems to be an object, but there is no object (or we cannot perceive the object), then we need it to be the case that there seems to be an object but it is not just the appearance of its properties. The misperception of its properties is a complex illusion.

So can we get a difference between the apparent object and the appearance of its complex of properties? Say all of the properties of the object are mistaken. Call this a maximally complex illusion. If a maximally complex illusion is not a hallucination, then what is the difference between them?

4.4.1. Property-Bearers

What is there in a complex illusion and not in a hallucination is what has the properties: In hallucinations, the bearer, the possessor, the instantiator of the properties is imperceptible. Let us call this the *property-bearer* of an object. One might then say this: The property-bearer is not a property involved in complex illusions; it is what bears properties involved in complex illusions. If there is no property-bearer, then we have a hallucination.

One may think of the object of hallucination—independent of the illusory properties—as the property-bearer. Alternatively, one may think of the object as constituted by the properties and property-bearer (e.g., the bearer is a bare substratum or bare particular (e.g., Loux 2002)). If this feature of an object (the property-bearer) is apparent beyond the appearance of the properties, then so long as the apparent property-bearer is identical to the actual property-bearer, we only have a complex illusion. If there is an appearance of a property-bearer which is not real, or which we cannot perceive, then this is a hallucination.

However, for this to work, one must already distinguish the appearance of the object from the appearance of its properties. Such an attempt seems to have a difficult task ahead of it. Its solution, however interesting, runs counter to the assumptions of the debate between substance theorists and bundle theorists. This may be correct to do so, but we cannot easily presume to do so.

Again, one reason the need for a property-bearer is a primarily metaphysical issue is that the phenomenology, how things seem, seems sufficiently captured by instances of properties at particular points in space and time. Phenomenologically, at least, so long as we include the space and the time, there seems to be no need for something other than the instances of properties at such locations.

Yet if this is right, then there is nothing to the appearance of an object over its properties—again, including spatial and temporal properties. There is no appearance of a property-bearer beyond that. So what distinguishes the complexity illusion from the hallucination?

One answer might be this: With complexity illusions, the properties are unreal or cannot be perceived (but the properties seem to be real). With hallucinations, the properties-bearers are unreal or cannot be perceived (irrespective of whether or not the property-bearers seem to be real).

Thus to establish the difference between hallucination and complex illusion, one must first show cases where not only are the properties neither real nor perceptible but also the bearers are neither real nor perceptible.

Note that this is a question about the structure of perceptual experience. It concerns whether, for an object, the bearer of its properties can be perceptually apparent, and not just the properties themselves. Otherwise, we are back to complex illusions.

Furthermore, this question requires engaging in contemporary metaphysical debate about properties. One must hold a metaphysical position in which it is the case that there is more to an object than its complex of properties (including, again, its spatial and temporal properties). This is not something one may assume metaphysically. It is one position in the debates in property theory; it is to endorse a substrate theory.

Such considerations might suggest a third option: give up on demanding that there is a difference in kind between illusions and hallucinations. Hallucinations are just particular kinds of illusions. They are complex illusions.

However, even if we do reduce hallucinations to a kind of illusion, they need not be complexity illusions as such. They can be another kind of illusion. This is an illusion of another kind of property, one related to certain collections of properties (and so certain kinds of complex illusions). I call these *sortal illusions*.

4.4.2. Sortal Illusions

The concept of a sortal illusion draws on illusions and conflationary errors. Indeed, given a certain understanding of the sort involved in a sortal illusion, the illusory component of it is a kind of complex illusion.

Consider the previous example of the Rubik's cube and the rubber ball. If I seem to see a Rubik's cube, but there is no Rubik's cube there, then I am mistaken about the appearance of a Rubik's cube.

However, I am not necessarily mistaken about the appearance of something. I do in fact see something: a rubber ball.

If I am not mistaken about the appearance of something, I am not mistaken about a bearer for the properties that are apparent to me. Thus, as defined earlier, I am not hallucinating.

This does not seem to be a hallucination but it also does not seem to be just an illusion of any arbitrary set of multiple properties. It is an error of a particular property or set of properties; I mistakenly identify the kind or *sort* of thing I perceive.

Many philosophers refer to the properties that we use to identify some object as the *sortal* properties of the object (e.g., Sider 2003; Hawley 2001). In this case, I have gotten the sortal property wrong, the property that I use to sort something into a kind or to place it under a category. I have gotten wrong the property that makes something seem to be the sort of thing it is—e.g., a ball, a piece of fruit, a cricket. In the earlier example, I have gotten wrong the property that makes something seem to be a Rubik's cube or a rubber ball.

We can distinguish sortal illusion from other kinds of erroneous experience as follows. Sortal illusion is distinct from

- An arbitrary complex illusion because the particular set of apparent properties places their possessor under a particular kind or sort.
- Conflationary errors because sortal illusions involve the appearance of particular properties, properties that identify it as falling under some kind. Conflationary errors need not involve the appearance of properties which place it under a kind; instead, they appear to be a multiple of particulars as one particular.
- Hallucination in the original sense because there is no requirement for there to be nothing there or perceptible. There can be a property-bearer beyond the illusory properties; it just does not have those properties and the related object is not that sort of thing.

This taxonomy of illusions and hallucination becomes important once we consider perceptual experience in light of different theories of time. Each kind of error involves different discrepancies between appearance and reality, in both degree and kind. Given how the actual world appears to us, this allows an easy of describing and evaluating different claims about the metaphysics of time in relation to experience.

4.4.3. Hallucinations of O Are No Instances of O

Given the introduction of conflationary errors, complex illusions and sortal illusions, can we eliminate hallucinations as a separate category or of error? Is it just a special kind of conflationary error or illusion (such as a complex or sortal illusion)?

Tempting to me as this may be, I think it is too quick. There is one last consideration. The reason why we need hallucination over illusion is that there are cases where, although there seem to be instances of properties, there is not just the absence of these instances of those properties (which gets us an illusion, complex, sortal, or otherwise). There are cases where there is either nothing perceptible or nothing at all. Yet there seems to be something perceived. There are no instances of properties, not even spatial instances or temporal instances. If we add in property-bearers, there are not even property-bearers.

This requires, with respect to the external world, that there is *nothing*. Alternatively, even if there is some object there, it is imperceptible. In that case, we cannot talk about illusory properties of some object 'out there' in the world.

To get to this, we need to show that we cannot perceive objects in the world. Then we might say that this is a case of a hallucination. Unlike complex or sortal illusions, one does not perceive anything; this is so whether or not 'anything' can be an object exhausted in its metaphysical structure by its properties or bare substrates. Unlike conflationary errors, we do not even have the instances of the properties, never mind their bearers.

To summarise, the idea of illusion and hallucination as distinct categories of erroneous experience can be problematic.

First, it is not clear what it means for an object to be apparent beyond its properties. So how can we set up a distinction between the appearance and the reality of the object, beyond the distinction between the appearance and reality of its properties?

Second, it is problematic if we do not endorse a currently debatable property theory, substrate theory, where there is a property-bearer beyond the instances of its properties. If we reject or are neutral towards that theory, a complex illusion or a sortal illusion seems sufficient for what we need for hallucination. Yet, if we make this identification, then hallucination is not a distinct category of error from illusion.

Yet we might have a concept of hallucination that can pick out experiences separate to illusions. In cases where one seems to perceive something, there is nothing that can be perceived.

Can we get such a case? More importantly, for some actual experience, can we prove that such a case applies to it? Can we show it or argue for it from the evidence available to us?

It may seem obvious that we can easily show it or argue for it. There seem to be many cases of just such a situation: people see strangers in their house where there is no one or even nothing there at all (e.g., Sacks 2010). Vivid experiences in sensory-deprivation tanks. Dreams. Given a proper understanding of the constitution and processes of perception and experience generally, we can show that there is in no way a possible of experience of something 'out there' in these cases. Thus, once we examine the evidence, it turns out that some, many, or even most of these are, in fact, hallucinations.

However, to show this, we must hold a position on what is 'out there' and what of it is perceptible. What is out there and what is perceptible depends on what can be real and spatially related to us. In later chapters, I argue that this depends on our metaphysics of time.

Before coming to that, it is important to discuss another class of error related to experience. This error is different to most of the errors introduced so far: general illusions, complex illusions, sortal illusions, and hallucinations. Conflationary errors are a form of it. We can think of it as a kind of blindness (or even double blindness). This is *anosognosia.*

4.5. Anosognosia

In my everyday life, I do not experience everything; everything there is not apparent to me. I do not even experience everything in my kitchen when I am in my kitchen.

For the most part, my everyday experience is this: I only experience some of the world in which I live. Only some of the world that surrounds me is apparent to me. Even the things closest to me in distance and scale—'medium-sized dry goods'—are not wholly experienced by me. Of an apple that I hold, I do not feel the entirety of its inside; I do not see the spaces between its atoms; when I bite into it, I do not taste, or even smell, and I certainly do not hear, the entirety of the inside of the apple. If I perceive the apple, I incompletely or only partly perceive the apple; I have a limited perception of the apple.

I have already discussed that it seems this way. A feature of typical experience is that there are perceptual limits; in having a perception, one in some way experiences something beyond what one perceives. This is amodal perception, a composite of perceiving and of some kind of experience that is not perception. There is what is perceptually apparent to us and what is imagined by us (if imagining is the right kind of experience involved).

First, our perceptions of the world are limited, partial, or incomplete. Second, our perceptual experience appears to be limited, partial, or incomplete.

In such cases, there can be an error. For example, there may not be a complete object, but only the part that is not amodally perceived (or something that resembles the part). There may only be a cat's tail with no hidden cat attached to it. Less grotesquely, there may only be an illustration of a cat's tail.

This kind of error does not occur with respect to what is not amodally experienced. The visible cattail is not mistaken; what is mistaken is that there is more of cat just out of view.

There is also another kind of error related to perceptual experience that is, in some ways, the opposite of amodality error. These are cases of experience in which we are not experiencing more than what is there. Instead, we are experiencing less than what is there and, through this limited experience, we are mistaken.

This kind of error is one in which, specific to a perceptual experience, one is unable to perceive everything. With visual experience, one is blind to what is there; with auditory experience, one is deaf (and so on, with other sensory modalities). In addition, towards these imperceptible entities, one is not aware *through* experience that one is limited, even by means of a complex experience encompassing the perceptual experience. For example, one is not aware through an accompanying amodal experience that one is blind, deaf, or numb to something.

The error is that one does not appear to have a limited experience of something when in fact one does. Here are some features related to this:

- *One perceives* something *of* O.
- *One has a perceptual experience of some O in a certain limited or partial way*. For example, there is an appearance of O that is F; there is no appearance of O that is G. However, O is F and O is G.
- *The limit itself is not apparent*. Continuing the earlier example, one appears to have an experience of O that is F but does not appear to have an experience that there is more than O being F. Yet there is; O is also G.

4.5.1. Experience Appears to Be Limited

If one is blind, one is unable to see; one's experience is not of the visible world. If one is deaf, one is unable to hear; one's experience is not of the audible world. Similarly for other senses: smell, taste, or even of one's own body.

Such inability can be partial. For example, one is blind in the left visual field or one has a scotoma (a blocked area of the visual field) in their right. One cannot hear out of the left ear but can the right; one cannot hear a difference in pitch between two notes when there is a difference between their pitches. One cannot distinguish the positions of a number of grains of dust or distant stars, or the different locations of the eye of a needle and the end of a thread. One cannot feel one's arm but can feel the rest of one's body; one cannot feel one's leg temporarily, for a limited time—e.g., after sitting still, or permanently.

Such inability can be normal and unsurprising (at least, as anything else). Everyone has a blind spot in the visual field, where nothing is visible. There is a minimal threshold to what can be spatially and temporally distinguished in vision and hearing. Certain rapid changes are not visible, nor are certain slow ones visible. No one human can see out the back of his or her head. No one can hear the full range of tones either between the extremes of pitches we do hear (and obviously not beyond them). We are deaf to the sound waves of stars. No one can hear conversation at the other side of the world. Similarly, for the other senses, I cannot distinguish the difference between one cell in my body and another, nor feel the differences between hairs on my head. I cannot distinguish the precise location of an infection when it begins, or the moment when my first hair fails to have colour in it, and I begin to grow grey. Nor can I smell a rat (unlike my cat).

In normal experience, experience is limited. That one's experiences, and what one has experienced, are only parts of the world is something I assume, and assume everyone reading this believes. It is one reason, for example, why it is intuitive to hold that our experience does not extend to the limits of the world (as discussed in the chapter on the structure of experience). It is why any apparent religious or numinous experience, where it is of everything in space and time, is extraordinary and strange. It is strange in comparison to typical human experience.

However, that does not mean we are always aware there are such limits. The psychologist Kahneman discusses in depth the way in which people make certain kinds of judgments. These judgments belong to a way of thinking in people that he calls System 1; there is also, in contrast, a way of thinking he calls System 2.

Importantly for this book, the judgments by System 1 are passed up to the more deliberative, abstract, conscious, slower, and critical System 2, which then evaluates or acts on these judgments. But in doing this, there is an implicit assumption which seems to guide the taking up of these judgments of System 1 by System 2. Unlike the judgments that System 2 itself makes, System 2 treats System 1 judgments as if they capture a sufficient amount of their target.

Kahneman calls this *WYSIAI*: 'What You See Is All There Is'. In an interview with the American Psychological Association, Kahneman states that, "WYSIATI means that we use the information we have as if it is the only information" (Kahneman 2012, 24). System 1 judgments are treated (in some way) as if they are based on sufficiently complete accounts of what is going on. There is no further enquiry necessary.

Compare using 'the information we have as if it is the only information' to amodal perception. With amodal perception, it is as if our overall experience is engaged in trying to get beyond the information provided only by perception itself. So we have an experience as of hidden elements, occluded objects partially seen (provided, perhaps, by our imagination). The information from perception is not the only information for us.

Yet some experience operates as if under a principle such as WYSIAI. In such experience, it seems as if all that we experience is all that there is. Yet this is false: In these experiences, they are still only of some of what we experience. It just does not seem that way.

Note that, since we have knowledge of the world in general, we may know otherwise, just as we may know otherwise that apparently bent lines are straight, as found in Hering illusions.

However, through the experience itself, for some of the things that we experience, there is no experience of otherwise. There is no amodal completion. There is no sense of some of what we perceive as only being partial. It is as if there is nothing more to what we perceive than what is perceptible.

This kind of error, where there is more to what we perceive than what is perceptible, I call anosognosia.

4.5.2. Defining Anosognosia

Anosognosia can be of anything we can perceive or perceptually experience, be it property, relation, or object (or, if one has perceptual experiences of them, sorts or kinds).

First, for property (or relation) F, S is anosognosic of O being F or being partly Q iff:

(i) S experiences O.
(ii) O is F.
(iii) F is not apparent.
(iv) S is unaware through the experience of O that O is F.

Second, for object Q that is part of O, S is anosognosic of F or O being F or being partly Q iff:

(i) S experiences O.
(ii) O is partly Q.
(iii) Q is not apparent.
(iv) S is unaware through the experience of O that O is partly Q.

O can be anything. For example, O could be the universe. It may seem that I experience all of it, a seemingly mystical experience, when I only experience part of it—e.g., the night sky. Or O could be something more mundane—e.g., a tomato, a chair, a cat. It can even be of oneself—e.g., one seems conscious of a simple undivided and persisting self, yet this is because one is unaware of a Parfitian complex self.

Again, F can be anything possessed by O, Q any part of O. F can be—e.g., a shape, a colour, or a relation to something else. It can even be of the perceiver's own properties—e.g., although one has some self-consciousness, one is also both unaware of aspects of one's self and that one is so unaware. Q can be the inside of an object, some concealed surface or side, some microscopic structure.

What is important is that, despite our experience of O, neither F nor Q are apparent to us. Further, we are not aware of that fact through our experience.

I assume that anosognosia can at least occur in all forms of perceiving: visual, auditory, etc. One can have visual anosognosia, auditory anosognosia, or tactile anosognosia.

Visual anosognosia is the main target here, as visual perception is our main concern. In addition, just like other kinds of erroneous experience, anosognosia need not be uncommon or surprising. What is required of it is that there is an error in having the experience.

We have already met a kind of anosognosia in conflationary errors. Conflationary errors are errors in which we identify one thing (or a few things) as having properties when in fact multiple (or more than a few) things have those properties. As such, the properties are not illusory nor are there hallucinatory objects. There are properties and objects bearing them. What is in error is *what* bears them; multiple (more than a few) bearers are conflated into one bearer (a few bearers).

Anosognosia can encapsulate this error. We do not perceive the differences between bearers of these properties, and so hold that they all belong to one thing (or too few things). This anosognosia of the multiple objects bearing properties that we actually perceive. This leads us to the erroneous judgement that there are fewer bearers of these properties than, in fact, there are.

There are many pathological cases, as evidenced by the label coming from many instances of such cases (e.g., Prigatano 2010). Anton's syndrome, for example, is a case where a sufferer is blind but does not know it, and insists that he or she can see. The person confabulates stories about what he or she sees, despite seeing nothing. There are two components to this: first, the confabulation and, second, the anosognosia. Although both may frequently come together in pathological cases, conceptually, confabulation is unnecessary for anosognosia; all that is needed is that the subject is unaware that he or she has limited vision. People may interpret this in a different way: they may interpret it as being the case that they do not perceive anything because there is nothing there to perceive.

The possibility I advance here is that, even in normal perception, one is unaware of some limits of one's perception. However, one judges this to be not because one does not perceive what is there, but because there is nothing there to perceive.

Let us take a case common in the literature on perception. We can understand this as visual anosognosia. It is also relevant later when I discuss temporal experience.

4.5.3. Change Blindness

Subjects, with normal vision, are unable to see a particular change; as a result, they conclude that there is not this particular change. This is change blindness. Examples are very easy to come by, both in media—e.g., pictures subtly or suddenly change without the change being noticed, and live—e.g., interviewers dramatically changing appearance or even who interviews (e.g., Blackmore 2003).

There are many conclusion one might draw from this error. Here is an uncontroversial one. Let us call the change to which subjects are blind C:

(a) For subjects when they report the experience, there is no visual appearance of C.
(b) For subjects when they report the experience, there is no awareness of a limit to vision that would exclude C.

That is, whether or not I do not see it at the time, when I report it, first, I do not remember seeing it. Second, I do not report what I see as if such a change could be hidden from me. That is, I am not aware that my visual perception is limited in this way.

4.5.4. Anosognosia Is Neither Illusion nor Hallucination

Anosognosia is not illusion or hallucination. With illusion and hallucination, there is more to what is apparent than is actually there. One perceives what seems to be O being F, but O is not F—e.g., because there is nothing (hallucination) or there is no F (illusion). With anosognosia, the interpretation of the appearance of O being F as mistaken is very different. With anosognosia, there is less to what is apparent than is actually there.

In cases of illusion or hallucination, a mistaken appearance of O being F does not give anosognosia of O being F. We are not blind, and do not know it, to O being F. O being F is apparent to us. That appearance is mistaken.

This distinction between, on the one hand, anosognosia and, on the other hand, illusion, and hallucination, opens up a way in which one might translate anosognosia to illusion or hallucination, or vice versa. This is dependent on how one understands the appearance of one object or property in terms of the non-appearance of another. I discuss this in the next chapter.

The difference between illusion (and hallucination) and anosognosia is not explicit in the literature on perception and error. The term anosognosia is not used explicitly this way because, as far as I am aware, it does not appear outside literature on pathologies and disorders. It has not been applied to typical or normal perception.

Yet the non-illusory, anosognosic way in which appearances differ from reality in pathological cases seems readily applicable to normal experience. It also corresponds to common situations such as change blindness. Writing about the relationship between the apparent location of sounds and the actual location of sounds, O'Callaghan writes,

> Another possibility is that we experience only a small subset of the locations sounds occupy during their lifetimes (for instance, while at

> their sources), and simply fail to experience where they are at other times. This avoids ascribing illusion.
>
> (O'Callaghan 2009)

Sounds might have multiple locations in their lifetime, with only a small subset of those locations apparent to us. This failure to experience the other locations is not a case of auditory illusion. However, it is the limited appearance of a sound's multiple locations—i.e., it is anosognosia of most of the sound's locations.

To summarise, there is a significant phenomenological and ontological difference between, on the one hand, illusion and hallucination and, on the other hand, anosognosia:

- *Illusion and hallucination*: One seems to perceive something (object or property) that is not there. One fails to perceive what is apparent.
- *Anosognosia*: One does not seem to perceive something that is there (and there is no experience of it). One fails to perceive what is not apparent.

There are a few final issues.

First, we can conceive of the unawareness in anosognosia in one of two ways: as being specific to experience or applying more broadly. In a broader understanding of it, this would be a case of our not even conceiving or thinking of O being more than F. The unawareness of O being more than F is the lack of any conception or reason to believe that O is more than F. This would be the case if we have no idea or concept of such a possibility, or if our theory disallows it. Bluntly, we just do not know and have no reason to think there is anything more than O being F.

I put aside this understanding of anosognosia. This is never the case where we have some reason to hold that there is more to O than its being F. As such, for there to be the anosognosia described here, the error must relate more intimately to the experience. Again, what is there is not apparent.

Second, one might argue that anosognosia as defined here is nothing new. Anosognosia is either just a limited perceptual experience, like all perceptual experiences, or an illusion of a particular property.

With respect to anosognosia just being an instance of limited experience, if it were true, I would be happy with this result. I would not need to introduce another kind of erroneous experience. I could simply interpret all alleged instances of it as merely limited experience. This certainly cuts down threats to naïve realism and, by extension, the untrustworthiness of experience. The only problem with this is, by means of amodal phenomenology, many limited perceptual experiences appear to be limited. Anosognosic experiences do not, and this is precisely why they are in error.

With respect to anosognosia being a special kind of illusion, this is more problematic. For what is the illusion? It seems to be an illusion of *no limits*, or of *totality*, or of *there being nothing more*. As the last phrase indicates, this is an illusion of something missing which is actually there—a limit, a particularity, something more.

If this is an illusion, it is one of negative properties. If one is happy with negative properties, and illusions of them, then this should be no problem. As discussed, I am not happy with them, but I leave that debate to rest here. For those who prefer illusions of negative properties, they should treat my terminology as referring to them as anosognosia of the positive counterparts (such as fullness) of their negative properties (such as emptiness). I hope that, beyond that particular difference, much of what I say here about anosognosia is unproblematic. (However, even in this case, as I argue in the next chapter, this status of anosognosia gives us reason to prefer commitment to it over other kinds of erroneous experience.)

4.6. Conclusion

This chapter has defined different kinds of erroneous experiences: various kinds of illusion, hallucination, and anosognosia. In detailing them, the taxonomy of erroneous experience has been expanded—notably, to include anosognosia, but also conflationary errors, complex illusions, and sortal illusions.

This taxonomy is an expansion from typical descriptions. As far as I know, much of what I say here is new. This may be a reason to be wary of it. However, as I discuss in later chapters, when trying to understand the structure of experiences in time (i.e., all experiences), it is important to determine the following:

(a) The kind of erroneous experiences that are possible
(b) The kind of erroneous experiences that are preferable in one's ontological commitments

While (a) is a theme of the whole book in that it is partially determined by what is true of time, (b) is relevant to (a), but can be considered independent of thinking about particular contents (such as space and time) or particular theories of time, and (b) is the subject of the next chapter.

Notes

1. For psychological discussion and examples of illusions, Lotto and Purves (2003). Also, an excellent online and academic resource for examples of optical illusions is Michael Bach's website (Bach n.d.).
2. In physics, such separation is commonly referred to 'space-like' separation.

3. I return to this in Chapter 5 when discussing choosing between errors and in Chapter 9 when discussing the experience of tense.
4. If bundle theory is true and all there is to an object is the compresence (or some other kind of bundling) of instances of properties, then we can explain the error as a mistaken identification of compresence (or other kind of bundling) of instances of such properties. I leave open how such bundling works or the relationship between such bundling and perceptual appearances.

5 Choosing Erroneous Experience

Should we explain mistaken appearances as illusion, hallucination, or anosognosia? One might think that there is no point in asking that question. In the process of identifying that there is an error at all, one classifies it as one kind or another. In identifying a mistake, one grasps also the kind of mistake there is. The kind of error is evident in there being an error.

However, there are situations in which we can choose the kind of error or even that there is no error at all. These are situations in which the kind of error depends on which account of reality is true, yet there is uncertainty as to which account should be accepted. Indeed, if each account provides a different kind of error (or even holds that nothing is wrong), it may be that a preference for one account is due just to this difference. One may choose theory T over theory T' because

(1) T results in fewer cases of erroneous experiences than T'.
(2) T commits one to a different *kind* of erroneous experience (e.g., illusion, hallucination, anosognosia) than T'.

For (1), I think the choice is uncontroversial or at least plausible. T puts us in a better epistemological situation than T'. T is a theory with less cases of erroneous experiences than T' and so is a theory under which experience is more reliable than under T', or even trustworthy in cases where T' is not.

However, for (2), it is not so obvious that one should prefer T over T', or T' over T. A preference of one theory over the other may only be a matter of taste. What other reason is there to prefer one theory over another based only on the kinds of experience it commits us to?

I reject one answer here at the outset: The best theories are the theories that commit us to the errors with which we are most familiar. For example, I am more familiar with thinking about illusions than anosognosia, so it is better to choose theories that commit me to illusions over theories that commit me to anosognosia. I assume that it is not self-evident that the less common kind of error is worse for a theory just in virtue of being less common. There must be at least something more said to defend why the less common kind is worse than its more common alternatives.

Another reason is more compelling. Any kind of error is less desirable if it involves increasing one's ontological commitment. For example, given the exact same phenomenology: Anosognosia is a case of there being more than there appears to be; illusion and hallucination are cases of there being less than there appears to be. As such, many theorists might be hesitant to prefer a theory with anosognosia to a theory with illusions and hallucinations. They might worry that it violates the principle of *parsimony* (at least more than theories with other kinds of error). Assuming that both theories meet what is otherwise necessary, why prefer the one that increases the ontology?

However, this is irrelevant to the work in this book. Yes, the competing errors of illusion and hallucination do not violate the principle of parsimony. However, they do violate another principle, the principle of *saving the appearances*. They both deny how things appear in a way that anosognosia does not. Anosognosia at worst denies that there is *only* how things appear; illusion and hallucination deny that there is *even* how things appear.

If we are empirically minded, this violation of appearances is worse than the violation of parsimony. It is also a worse epistemological situation to be in—cases in which illusions and hallucinations undermine our experience in a way that anosognosia does not. Theories that commit us to anosognosia place us in a more positive epistemological situation than theories that commit us to illusion or hallucination.

We are in a position to choose between errors if we can choose ontologies that have different relationships to appearances. We are in this position when what differentiates one kind of error from another is not appearances, nor is it the relationship between appearance and reality. It concerns differences in what each error demands of reality. We are in that position whenever we can choose a side on the metaphysics of time—to be eternalists or presentists, to be tense theorists instead of tenseless theorists.

Yet perhaps we can choose between errors without changing our ontological commitments.

5.1. Ranking Errors for Preference

Could we perhaps choose between errors without changing ontological commitments? For example, could we choose to interpret an erroneous experience as anosognosic instead of hallucinatory, or vice versa, yet remain committed to the same position about what is real in the world?

5.1.1. Interchanging Illusion and Anosognosia

Anosognosia is not illusion, and illusion is not anosognosia. However, there are possible cases in which we can re-interpret the two kinds of errors in terms of each other without ontological consequence. That is, we can re-describe a discrepancy between appearances and reality without having to change our commitments to what is real.

A case of anosognosia can be a case of illusion, and vice versa, so long as both are of different properties related as follows: the non-appearance of one property is the appearance of another. In this case, it is an ontologically neutral choice whether or not one wishes to describe an experience as illusory or anosognosic.

Anosognosia of some property F is an experience in which there is F and the *non-appearance* of F. To get an illusion, we need, if there is a non-appearance of F, then there is the appearance of something else. Call it G. Then

(a) The non-appearance of F is the appearance of G (as already stated).
(b) If there is F, then there is not (G). If there is G, then there is not (F).

If (a) and (b), then a case of anosognosia of F (the non-appearance of F) gives a case of the appearance of G but, really, not (G). That is, it gives us a case of an *illusion* of G. Thus anosognosia of F gives an illusion of G.

Let us consider some examples. There are two ways we can interpret them:

- As cases of both illusion and anosognosia: The property in anosognosia and the property in the illusion are incompatible. However, neither is reducible to a mere absence of the other.
- As cases only of anosognosia: The 'illusory property' is merely the non-appearance of another property.

(a) The Non-spatiality of Mental Events

Consider the properties of *being far away* and *being nearby*. If anything, far away (F) is not nearby (not (G)) and anything nearby (G) is not far away (not (F)), then what seems nearby (apparently (G)) does not seem to be far away (not apparently (F)). If what seems nearby is, in fact, far away, then we have (a) anosognosia of it being far away (it does not seem far away, yet is) and (b) illusion of it being nearby (it seems nearby, but is not).

Yet this only holds if it meets an important condition: The absence of one property is the *presence* of some other property. In the earlier example, that something is *not far away* means that something is *nearby*.

Experiences need not always meet this condition. The absence (or non-appearance) of a colour, for example, is not obviously the presence (or appearance) of another colour. Sacks discusses the case of a painter who, after an accident, lost the ability to see any colours at all (Sacks 1996). One might say the painter is not experiencing colour if he is unaware that he lacks this feature.[1] However, he is not then under an illusion of *some other colour*. He is not aware of any colour at all.

Another example comes from a common dualist thought that mental states have no spatial location. Say such mental states appear[2] neither far away nor nearby. If they do not appear that way, it is not because there is a failure to perceive them where they *really* are. They neither appear to be nor are spatially located. They are nowhere. This is not a case of illusion or of anosognosia.

Alternatively, mental states are indeed spatially located. Yet they do not seem to be spatially located. Again, this need not be an illusion. We fail to perceive them as being where they are, in fact, located. We are anosognosic of their spatial location and do not know it. However, we are not under an illusion of something else.

The last example may also do for another approach: We eliminate one kind of erroneous experience in favour of another. In this case, we only have the absence of one property without commitment to that absence being another property.

For example, with mental entities, we may eliminate an illusion of non-spatial location as follows:

- There is no appearance of spatial location in the appearance of mental entities.
- One is able to argue successfully that this non-appearance of spatial location is not the *appearance* of spatial *non*-location for mental entities (whatever that might mean). It is just the lack of appearance of spatial location.
- If the mental entity has no spatial location, there is no error here. However, if the mental entity has spatial location (e.g., in the brain), we are anosognosic of that spatial location.
- If we are anosognosic, we need not say that we have an illusion of spatial non-location as well. This is not something that is *apparent*. It is just the lack of something apparent.

If one previously understood the erroneous experience about mental entities as illusions of non-spatiality, one can put aside this interpretation. The error is anosognosia. The spatial locations of mental entities are imperceptible; we do not know this. Thus we erroneously judge that mental entities have no spatial location. Yet this is not because mental entities illusorily *seem* to have a property of spatial non-location.

(b) The Self-Motion of a Magic Flower

I described a magic trick in the last chapter. A flower appears to float from a magician's hand and across the stage by its own volition. Is this an illusion or anosognosia? To answer this (just as the authors of the book do from which this example comes) I must ruin the trick.

The magician is not the only person on stage. There are also accomplices dressed entirely in black (the same colour as the curtains). When the light shines on the stage, it brings up extremes in illumination. The white objects are much brighter than the black objects behind them. Our eyes adjust to such bright light by cutting down on degrees of contrast of the range of illumination we can detect (just as they do when we suddenly step into sunlight from a dark room).

As such, differences in colour and illumination between the much darker objects become indistinguishable. We cannot detect the difference between the accomplices and the curtain behind them. We are so completely blind to such low-illumination differences that we do not even notice the accomplices' movement. (Furthermore, there is music playing so loud that we cannot hear the accomplices move about.) When an accomplice picks the flower out of the magician's hand and walks across the stage with it, we neither see nor hear them walking. We do not see them carrying it. We only see the flower.

We also do not know that there is this failure in our visual experience. Further, our experiences are not amodally completed. We are not having an experience that fills in gaps in what is happening in the darkness and behind the noise.

There is an erroneous experience here. However, is this magic trick an illusion or anosognosia?

First, the flower appears to move across the stage. This is a veridical appearance. The flower *does* move across the stage. This motion is not illusory. What is illusory is that it moves on its *own*. It does not move on its own (someone carries it).

One may call this apparent motion an illusion of *self*-motion. However, it is a limited perception of *aided* motion accompanied by ignorance of that limit. It looks to be self-moving only because nothing else appears to move it. This is anosognosia of *aided* motion.

We might decide that both self-motion and aided motion are both properties that things can have. However, they are the opposite of one another. If something self-moves, then it is unaided; if something is under aided motion, then it is not self-motion.

As such, we may decide also to hold that there is both an illusion of self-motion and anosognosia of aided motion. An illusion of self-motion is not only a failure of inability to notice what actually causes the motion. Alternatively, we may decide to hold that there is no actual illusion here. An illusion of self-motion is only anosognosia of aided motion. There is nothing more to an 'illusion of self-motion' to require that we explain the appearances beyond what is explained by anosognosia. If we can have the latter case, then we need only commit to anosognosia here.

5.1.2. Interchanging Hallucination and Anosognosia

Can one re-interpret hallucination as anosognosia (and vice versa) in the way one might for illusion? Not intuitively.

Hallucination involves there being no objects (or at least no objects that can constitute experience). Anosognosia involves there *appearing* to be less (less objects, less properties) than there is (or that constitutes experience). If we can interpret a hallucination as a case of anosognosia, we can re-interpret a case of no objects as a case of more objects than there appear to be. This would be an appearance of less than no objects. Whatever this means (I cannot say for sure), this is not how hallucinations appear. To a hallucinating subject, there appears to be of at least one object (not less than no object).

Instead, to re-interpret as anosognosia an erroneous experience described as a hallucination, one must say something about what is there. In so doing, one must *deny* hallucination. It is not that hallucinations are anosognosia described in a different way. Anosognosia involve commitments to very different things.

As such, one might try the other option: one *eliminates* hallucinations in favour of anosognosia. Yet, even then, without changing one's metaphysical position, one cannot simply eliminate it. This is because, again, one has to eliminate not only the appearance of some non-property (by interpreting the situation as the non-appearance of some property). One has to eliminate the appearance of an object.

Another way to get anosognosia from an alleged case of hallucination is by having it that there is no error in the appearance of an *object*. That is, there is (at least) an *object* there and we perceive it. What goes wrong is that the particular object that appears to be there is not there. It is a different object. The mistake is in the properties it appears to have, properties that make it appear as if it is something else entirely. That is, we re-interpret the alleged hallucination as an illusion. We then get anosognosia from the illusion.

However, to do this, we must change our ontological commitments. With hallucination, we deny that there is an object (or at least that there is an object constituting the experience). With illusion, we accept that there is an object (that constitutes the experience). The mistake is only in the properties.

5.1.3. Interchanging Hallucination and Illusion

In previous chapters, I argued that one could have evidence of complex or sortal illusions that might appear to be hallucinations.

Strictly speaking, this interpretation of evidence for a hallucination as evidence for complex or sortal illusions is not a re-interpretation of hallucination as illusion. It is a re-interpretation of the *evidence* of hallucination to *evidence* of illusion. In changing from hallucination to complex or sortal illusions, one is not simply interpreting the evidence without ontological consequences. One has an object or property-bearer in complex and sortal illusions; it just does not have those complex or sortal properties it appears to have. One has no object or property-bearer in hallucination; there is nothing there at all with properties or not.

However, the difficulty with this strict difference is that 'the object' that is there in complex illusions but not in hallucinations is either one of two kinds:

- A particular property-bearer without properties—i.e., a bare particular or substratum.
- A particular property-bearer with properties—i.e. a concrete particular. It is just that the properties are wholly different to those of the apparent object.

The problem with these kind of objects is that they seem to be as likely to be there in hallucinations as in illusions. Some object with properties other than the apparent properties are likely to be there even in hallucinations. The world in which subjects and their experiences are embedded are not voids. It is filled with things, perceptible or imperceptible. There is always something there. Whatever is there need not have any of the properties of the allegedly hallucinated thing (or even any properties at all—i.e., it is a bare particular, although that is a much less compelling claim). Yet that is the same for a complex illusion. In a complex illusion, whatever is there need not have any of the apparent properties of the object.

Thus to separate hallucinations and complex illusions, we need to come up with a reason for why the object in complex illusions is a constituent of the perceptual experience (it is just that its properties are not) but the object in hallucinations is not a constituent. If we have that, then we can hold that hallucinations and complex illusions are different. If we cannot have that, then it seems we can re-interpret hallucinations as complex illusions.

5.2. The Preferable Error

We have three possible kinds of erroneous experience: illusion, hallucination, or anosognosia. We might interpret at least two kinds in terms of the others. We can conceivably re-interpret any evidence of one in terms of the other, provided we make some change in our ontological commitments. Yet why would we re-interpret one kind of error in terms of another?

One reason might be a kind of parsimony about error. We wish to cut down our commitments to the number of error kinds. We wish to re-interpret all possible cases of one kind of error into cases of another—e.g., re-interpret all cases of anosognosia as illusions.

From an epistemological perspective, this reason is insufficient. If we reduce our commitments to one kind of error, that kind may yet be so epistemologically problematic that perceptual experience is greatly unreliable and untrustworthy. We may find that perceptual experience goes wrong in only one particular way. However, the way it goes wrong is so common and wide-ranging that we are always in a position of uncertainty as to whether or not the perceptual experience shows us anything about the actual world.

A better epistemological reason is that we commit to the kind of error that involves the least difference between erroneous experience and trustworthy experience. That is, many of the commitments required to hold that there is such a kind of erroneous experience are the commitments required for trustworthy experience as well. A particular erroneous experience is still a mistaken experience, of course; however, it is mistaken in a way that, other than that particular experience, it gives no reason to hold that perceptual experience is untrustworthy.

In that case, I think that anosognosia should be preferred to illusion or hallucination. This might not be obvious.

Anosognosia requires ontological commitments that illusions and hallucinations do not require. Anosognosia needs there to be more there than there appears to be. In contrast, both illusion and hallucination need it to be the case that there appears to be more than what there is (or, if there is at least as much there, what is there is not part of the experience).

As such, given the same phenomenology, illusion and hallucination involve less ontological commitment than anosognosia. Furthermore, hallucination involves less commitment than illusion. For example, if I hallucinate a divine being, there need not be

- A divine being—as needed by trustworthy perceptual experience;
- Something with non-divine properties the incomplete perception of which is an appearance of divinity—as needed by anosognosia;
- Something with merely apparent divinity—as needed by illusion; and
- Anything at all.

One might think that, combined with the principle of parsimony, such a difference in ontological commitment makes hallucination preferable to anosognosia and illusion, and illusion preferable to anosognosia. The principle of parsimony is to avoid multiplying the number of entities beyond necessity. Let us assume that appearances are necessary. In that case, anosognosia multiplies the number of entities beyond necessity. It involves claiming that there is more than the appearances, and much of that is contingent, not only necessary. Hallucination and illusion do not; there is less than necessity.

However, if we take appearances to be necessary (we may not do so but then this relationship between erroneous kinds is irrelevant), then illusion and hallucination are not better than anosognosia. It may not be best practice to avoid adding to what is necessary; however, it is at least coherent. There may be more than what is necessary (in this case, what is apparent). It is worse to deny what is necessary. That is to be incoherent. In this context of taking appearances to be necessary, by denying appearances, illusion and hallucination deny what is necessary.

This raises a different question: Is it better to add to appearances or deny appearances? That is, is it better to have more than there appears to be—i.e., as with anosognosia—or less than there appears to be—i.e., as with illusion or hallucination?

The issue is really about whether we want to save or deny the appearances.

I think, where these options face us, we should choose to save the appearances. To adopt a principle of *denying* appearances (of always having less than appearances) is not obviously a parsimonious one (such as *Occam's razo*r). It is a principle where one cannot match appearances and then commit to *less* than what is apparent. One cuts away appearances (it is more like an *Occam's Cleaver*). This is what we get when we choose illusions and hallucinations over anosognosia, or hallucinations over illusions. We choose to deny appearances.

Note that this is only an issue where we can choose the kind of error. If we must have illusions and hallucinations, then there is nothing we can do. What is apparent and illusory or hallucinatory is just not a part of reality.

Another reason we may choose to posit there being more than appearances is that it is plausible to assume that perceptual appearances do not exhaust reality. What is perceptually apparent is only a part of what there is. Only some of the real and actual world constitutes are perceptual experience. Experience is embedded in the wider world, extended over only a fraction of that world.

Anosognosia requires no more than that. The error of anosognosia is that this additional part of the world is not perceptually apparent to us, leading to mistaken judgments about the world. In contrast, illusion and hallucination require that our perceptual capacities can generate the appearance of something (property or object) that is not part of the world at all.

Anosognosia involves commitment to some thing to which even the most naïve theories of perception and cognition are already committed. It involves commitment only to limited perception and to ignorance. Anosognosia is ignorance of one's limited perception, of one's inability to perceive what is there. Given some A, one is anosognosic of A if one does not perceive A and one does not know that one does not perceive A.

Conceived as a kind of error, anosognosia requires no further explanation beyond the explanations of ignorance and limited perception. Both are general features of cognition and consciousness before any theorising about perception. Whatever else we commit to, these are both ways in which experience must be.

Whatever we say about illusion and hallucination, we cannot as easily reduce them to generally accepted features of cognition and experience. They do not just involve limitations of perception and ignorance of those limits. They involve a claim that experiences are phenomenologically richer than what is there or is perceptible.

To see the difference, consider the following example of a mistaken appearance of an object's shape. I see a three-dimensional cylinder only illuminated from one side. It appears to be a two-dimensional rectangle. Let us say I believe that or it seems to me that I am seeing the entirety of the object—i.e., that it is only rectangular, it is only two-dimensional, and nothing more. Of course, I am mistaken; there is more to the cylinder than what is visible; it is three dimensional and cylindrical. Is this the way in which I am mistaken an illusion of its rectangular shape?

No. In the two-dimensional plane containing that particular side, the object is rectangular. This is not an illusion of that spatial property. What the cylinder does not have is *only* that spatial property—i.e., a two-dimensional spatial property. My mistake is to take what I perceive of it to be all there is to it. My mistake is to identify the cylinder's perceptually apparent properties with all of its properties, its apparent two-dimensional shape with all of its shape.

In contrast, a different situation exemplifies illusion: I see a one-dimensional line. It appears to be a two-dimensional rectangle. In this case, the apparent shape (a rectangle) is more than the actual shape (a line). My mistake is not only to take what I perceive of it to be as all there is to it—i.e., that it is only rectangular and nothing more. The line is not rectangular. What is perceptually apparent about the line (a rectangular shape) is more than what I can perceive of the line (neither the line nor a part of the line has a rectangular shape). My mistake is not only to identify the line's perceptually apparent properties with all of its properties, its apparent two-dimensional shape with all of its shape. I am not seeing anything about the line that it has. It does not occupy two dimensions. There is no way to encounter the line in such a way that it appears to occupy two dimensions.[3]

Thus since they only appeal to generally accepted features of cognition and experience, theories that commit one to anosognosia are preferable to theories that commit one to either illusion or hallucination. We can see this in how theorists appealing to each kind of experience can respond to objections from those who hold a perceptual experience to be veridical.

Let us say you make a claim about the actual world that there are certain things in it—e.g., physical things. Further, you say that these are the only things in it—e.g., you hold that there are *only* physical things (i.e., you are a physicalist).

As already discussed, we might test this claim by how it corresponds to appearances. Naïve theorists, in particular, use appearances to justify and constrain a theory of reality, because they hold appearances to be (at least part of) reality. In this example, you test it by seeing if there only appears to be physical things.

Let us say a naïve theorist objects to your claim based on appearances. How things seem is not how you claim them to be—e.g., there do not appear to be only physical things. From the perspective of saving the appearances, they have the advantage; they save the appearances by accepting them as given (it is perhaps more correct to say the appearances save *them*).

Given anosognosia, you accommodate the naïve realist's appearances as follows: those appearances are incomplete experiences of what you hold there to be. There may appear to be some non-physical things. However, that is because there are more physical things than there appear to be, and those other physical things explain the appearance of non-physical things. You just cannot experience all of them. Instead, what you experience is a subset of physical things.

I assume that the objector accepts that there are limits to experience and that one can be wrong about such limits. What an objector needs to argue against claims of anosognosia is one of two things: first, that, on this occasion, there is *not* more than there appears to be or, second, that the 'more' is not what the defender of anosognosia claims it to be.

Both move the debate away from the appearances into the reality beyond them. Both naïve theorists and defenders of anosognosia no longer debate the correspondence of appearance with reality to do with the specific content. They debate about whether there is anything beyond that content.

However, if one appeals to illusion or hallucination, one accommodates these appearances as follows. They are not merely limited experiences of what is there. How things appear is not due to something missing. There is a fundamental error in experience in that there seems to be more than is there. It is not possible to put aside the appearances. The dispute is over the veracity of appearances, with an uncertainty about whether we can save the appearances.

5.2.1. *Illusions of Totality*

One might raise the following exception or counter-example to this analysis of anosognosia as a non-illusory (and non-hallucinatory) error.

When I am anosognosic, I do not just believe that what I perceive is all that there is. It is apparent to me that what I perceive is all that there is. There is the appearance of exhaustion or totality of some class of this content. For example, I see all the colours something has, or the whole shape, or everything that moves. As such, one might claim that I have something apparent which I cannot explain as merely ignorance of my limits. There is a phenomenology of (say) totality to it. The mistaken phenomenology is an illusion of totality.

There may be a phenomenology of totality, and so, in this case, an illusion. Even so, this particular mistaken appearance still makes anosognosia the better option. What is mistaken is a single apparent feature—one shared by all failed experiences of this kind. One has only one illusion in play: the illusion of totality. Regarding particular experiences, one suffers anosognosia so long as the appearance of that content is of *all that there is* of that content. This is an erroneous experience of nothing more to that content—e.g., to colour, shape, individual object, etc. This is the case even if it is an illusion of this property for objects, properties, relations, or anything else.

In any case, one might wonder if it is correct to conceive of this as an illusion of totality, as of a particular kind of *content*. This is appearances of totality separate to other appearances, for example, to the colour, shape, or individual object. We can ask of the objector, if we account for the appearances of certain properties, relations, and objects, what more does an appearance of totality add to that?

One answer is that illusions of totality are the *non-appearance* of something *more* of other content—e.g., of the colours, shapes, individual objects. That is, the illusion is of a kind of negative property. I have already discussed the concept of the appearance of negative properties. In addition, recall the discussion on amodal perception. In amodal perception, imagination (if that is what does it) marks perceptual limits, supplementing perceptual experience with what cannot be, according to naïve realism, a constituent of the perceptual experience itself.

If this is right, then I suggest the following re-interpretation of the phenomenology involved in 'illusions of totality'. This is a perceptual experience in the absence of such supplementary amodal imagination. This gives the appearance of totality: one seems to perceive everything because what we perceive seems to be all that there is. If our perceptual experience is supplemented by imagination (e.g., as how some theorists explain amodal perception), then one would not seem to perceive everything. It would not seem to be all that there is.

That is, this 'appearance of completeness' is due to the absence of the imaginative capacity found in other perceptual experiences.[4] This makes it similar to limited perception and the lack of belief. It is due to an absent, limited, or failed amodality. This absence gives rise to anosognosia.

There is more to say here about the appearance of totality. In a later chapter, I consider erroneous experiences of absoluteness when we encounter relative properties.

5.4. Conclusion

Say that we can take some kinds of erroneous experience and re-interpret them as or replace them by other kinds of erroneous experiences; say also that we have reasons to prefer some kinds to others. Then this suggests an approach to error in perceptual theory: We take instances of the less desirable kinds of erroneous experience and look for ways to re-interpret them or replace them by the more desirable kinds of erroneous experience. In doing so, we may have to change our ontological commitments and so choose between different theories about what is real or actual in the world.

This is not to say that we must do this or that we must prefer anosognosia to other kinds of errors. We may still prefer hallucination or illusion to anosognosia, where hallucination and/or illusion are compatible with what must be there and anosognosia is not.

However, in that case, anosognosia just fails to capture the ontology. It involves a false view of what is there. Anosognosia is not available to explain the discrepancy between appearance and reality because it requires unacceptable claims about reality. If one has good reason, otherwise, to deny that there is an x there, then one cannot say that one is anosognosic of x.

To return to the example of divine visitations in my childhood: I may explain the appearance of moving statues by appealing to their movement by human-sized creatures with magical powers. We are just anosognosic of these creatures or the magic they use to move the statues. These creatures and their powers are undetectable. However, if there cannot be undetectable human-sized creatures with such abilities, appeals to anosognosia of them or their actions is irrelevant.

To interpret experience as erroneous in one of these ways and not the other comes down to one's ontological commitments. Yet a general preference for some theories over others based on their commitments to different kinds of error can also influence one's decisions to make such ontological commitments. This brings us to time. Different positions on time involve different commitments to what is real. These different commitments include different commitments to the structure of perceptual experience, including erroneous perceptual experience.

Notes

1. If *colour eliminativists* (e.g., Arstila 2003) are right, then such a subject is not under any erroneous perception at all. The world is not coloured; it does not seem to be coloured. How things seem is how things are.
2. One might take this to be a kind of cognitive experience or cognitive phenomenology. I have not discussed this kind of experience.
3. I give this example assuming that there is no way to encounter the line in such a way that it appears to occupy two dimensions. When I discuss mirages in later chapters, I argue that there is a way we can encounter something line-like that mistakenly appears to occupy two-dimensions (e.g., the stick in the 'bent stick' illusion)—and yet it is not illusory but anosognosic.
4. Why no amodality in this experience but amodality in others? I take that to be a question about perceptual contingencies, answered by such approaches as evolutionary and development psychology—e.g., perhaps we are anosognosic of properties that our ancestor's survival did not require awareness about, e.g., the duration between a seen supernova's glow and the subjective activity of seeing the glow. In contrast, where it aided survival, amodality compensates for such perceptual limits.

6 Spatial Relations to the Past

> The dead appear to us in dreams, said Julian, because that's the only way they can make us see them; what we see is only a projection, beamed from a great distance, light shining at us from a dead star.
>
> (Donna Tartt, *A Secret History*)

Some of the erroneous experiences discussed in previous chapters are errors about depth. Given the theme of this book, one might wonder if these errors are illusions or anosognosia of depth.

In the next few chapters, I discuss the relatively uncontroversial claim that there are illusions of depth. What makes them illusions is that depth is an apparent property of what we perceptually experience. I argue that the appearance of depth suggests that there must be illusions of depth. However, the precise nature of these illusions—what exactly goes wrong—depends on which philosophical position about time is true.

The reason the error depends on which philosophical position about time is true is that it depends on what can constitute our perceptual experience. What can constitute our perceptual experience depends on which philosophical position on time is true. This is partly due to how we can be spatially related to what we perceptually experience.

When I go out at night and look up, it seems to me that I see a sky full of stars. All of these stars shine sharp and bright in the sky, as real as my hand shines in the moonlight. I cannot tell exactly how far they are, but I can say this: it is very difficult for me to accept that what I am experiencing as I look up is not something up there, but in my head. It is also very difficult for me to accept that what I am seeing when I look up is *not real*. Those stars, I want to say, are real and out there, just like the rest of the world. I know this because, every time I look up, I visually experience—or see—them *up there* (wherever that 'there' might be).

This would suggest that such stars are as ideal candidates for constituents of my perceptual experience as something closer to me—e.g., when I look at it, the surface of my hand. That is, at least as it *seems* to me, such stars are entities that seem to be real *and* perceptible.

Unfortunately, some concepts of time deny that these stars are apparently external, and real stars can be real and perceptible. Indeed, these concepts of time deny that, strictly speaking, I see anything else that is external and real. The concepts of time that do this are precisely the ones said to be the intuitive or folk concepts of time. That is, our ordinary everyday intuitions about time conflict with our ordinary everyday intuitions about what we can see.

The problem with the intuitive concepts of time has two parts:

First, these concepts generate problems concerning our spatial relations to external things and thus any phenomenology of spatial relations to such things.

Second, they generate problems concerning the reality of external things and thus our phenomenology of the reality of such things.

6.1. Naïve Commitments from the Phenomenology of Perceived Space

As far as I can tell, I am able to see things around me. I can see the external world. Along with the stars, I can see people outside along the street and the cup of coffee on the table before me. Not only can I see these but also I do so, again as far as I can tell, immediately or directly. They are just there around me, before me, visible when I turn my head or glance around.[1]

Some of them are at a distance from me. Not all are visible together. Further, my *understanding* of what I see at least partially depends on what I have learned in the past; my understanding of the world I perceptually experience does not come entirely from my perceptual experience of the world. Yet, as far as I can tell, I do perceptually experience some of it, this actual world. In addition, what I experience is not only what I already know. My experience does not seem to be an echo of my thoughts, not even a complex echo, not even of many thoughts. When I move around, I experience regions of it here and there, depending on where I go, and some of those regions contain things I never knew or thought about before.

Not only do I see things in the external world, I feel and hear and smell some things in the external world as well—e.g., the wind on my arm, the traffic on the main road, the coffee in the cup. In many of these latter cases, it can often be far more difficult to identify what I perceive, in terms of both what the thing is and its properties. I am not as familiar with the sight of a song thrush as I am with the sound of a song thrush. I do not know from which direction the wind along my back comes. I have different degrees or amounts of knowledge about what is apparent to me under different sensory modes.

Yet whatever the things that I smell, touch, see, hear (or, indeed, however these are fused together in a multi-sensory experience), what I perceptually experience appears to be outside and not merely subjective activity. It is independent of me, of what goes on in me; it is also independent of what goes on in my mind (extended or embedded), of my sensory organs, or of anything else that can be said to be mine.

In addition to apparent externality (and independence from my mind or myself), any events or objects that I perceive, again, as far as I can tell, are *real*. They are not imaginary or dependent on mere belief and fantasy; they are not even dependent on knowledge or perception. My knowledge or perception connects me to them in some way, but does not determine them. I detect these things and their properties; I do not construct them. If I were to cease to exist, then as far as it appears, what I seem to perceive (e.g., see, touch, taste) would still exist.

This is a *naïve* description of my perceptual experience. If my beliefs correspond to that description, then we can call them naïve beliefs. The description is naïve because it comes only from how things appear to me. I am drawing only on the phenomenology of the perceptual experience, on the perceptual *appearances*. Although I qualify it with reference to other cognitive states, showing that I can have further complex relations to what is perceptually apparent, this is not a reduction, elimination or denial of this phenomenology.

In particular, to hold that what is perceptually apparent to me is real because of how it appears is to be a naïve realist about perceptual experience. Holding that such phenomenological descriptions are accurate commits one to a number of positions about perception:

(1) *Reality*: What I perceive is real, unlike imaginary or depicted things.
(2) *Externality*: For the most part, what I perceive is *outside* of both any of my mental activity and me as a subject.

6.1.1. Externality

'Externality' can have one or both of two meanings. It can mean (a) 'mind-independence' or (b) 'physically external'. The distinction between these meanings is important: From accepting that something perceptually experienced is external, there are different commitments from (a) and (b) to the constitution of perceptual experience.

(a) Mind-Independence

My mind does not encapsulate the entirety of what I perceive. I could stop thinking about what I perceive, there could be no mental activity at all related to them, and they would exist; they would be real. Phenomenologically, such things include shapes, motions, and distances from my body's location, and their colour. (Again, this is how it appears. It need not be how it is).

However, one can make a further ontologically significant distinction in the understanding of mind-independence itself.

A. If something is mind-independent, then mental entities—e.g., thoughts—need not constitute them. Something mind-dependent must be constituted by mental entities (even if only partially).

Complex thoughts are an example of a mind-dependent entity. For example, the complex thought P, 'I had a nice dinner but a bad sleep', is constitutionally dependent on thoughts about 'a nice dinner' and 'a bad sleep'. Remove either of these constituting thoughts and remove P. At best, only the other parts of the thought remain—e.g., take away 'nice dinners' and only 'bad sleeps' remains.

We might still conceive of such thoughts, although complex, as real (as (1)). However, they are not external to the mind (against (2)).

B. They are not merely intentional entities. This is unlike unicorns, or my winning the Nobel Prize in philosophy. One might conceive of such things as external to me, but they are not really external. As they are not real, they are not really anything. There is no such medal for philosophy, so an event including such a medal is not actually internal or external.

Again, this is because in some way, one might say that they are mind-dependent or internal, as many philosophers do. However, this can be a little misleading, for they do not exist. It is more that there are only mind-dependent entities that refer to such things—e.g., descriptions, definitions, or *ideas* of them. Merely intentional entities are nothing at all in the actual world.

(b) What I Perceive Is Outside of Me as a Spatially Located Entity

The things that I perceive are physically outside of me, where 'me' refers to my body as a whole or parts of my body (e.g., such as my brain). Except for certain experiences such as pains, much of what I experience is external in this way.

Thus when I say that what one seems to perceive is *external*, I mean that what one seems to perceive is not only one or more of the following: (i) necessarily mental, (ii) constituted by mental entities, (iii) only intentional and irreal content of thought, and (iv) within the perceiver's body and brain.

Thus this is how it appears to me. This is the phenomenology: I perceive real external things. If this is how it seems, then why not have it that way? Why not be a *naïve realist* about this? Why not take how things appear to be the same as how things are? Why not take the *phenomenology of* my perceptual experience to be identical to the ontology of my perceptual experience? Why not include the entities that perceptually appear in the structure of perception?

The answer may be obvious: *We have illusions and hallucinations of external things*. All of these appearances are possible without there being anything in the world corresponding to them. As such, in perception, we cannot infer from how things seem to how things are.

Philosophers might think that they have a ready reply to this. *Disjunctivists* hold that one can (and should) distinguish perceptions from illusions/hallucinations and that there is no reason to infer from the latter anything about the former. Disjunctivism labels illusions/hallucinations 'bad cases' and perceptions 'good cases', then denies that there is anything in common—no 'common kind'—between the good cases and bad cases.

I have already raised one epistemological problem with this sort of disjunctivism (see the introduction). Another problem is this: Given a certain way of thinking about things in time, for all typical alleged 'good' cases of perceiving external things, there is a reason to deny the appearance of external things and to deny the veridicality of that alleged 'good' perception. If that is right, under disjunctivism, we should label all cases of the perceptual experience of external things as bad cases.

This is because we cannot perceive external things. We cannot because of a fact about the physical world. There is a time lag in perception. At least, we cannot if we assume alleged intuitive concepts of time.

The problem turns on an intuitive answer to the following two questions:

- Can non-present things and events be spatially related to me?
- Can I perceive non-present things and events?

For both questions, the answer from alleged intuitions of time is *no*. We cannot be spatially related to nor perceive non-present things. If this is right, then we cannot perceive non-present things that are spatially related to us.

Coupled with time lag, this means that we cannot perceive the numerous things we seem to perceive.

6.2. Directly Perceiving the Present

I mention seeing stars. I do not see stars right now. I see the café I am in and the cup of coffee before me. I feel buses and traffic rumble by outside. I hear music and the clanking slam of the waiter making coffee and using the till. In the actual world, for those of us whom empirical evidence partly determines our beliefs, there is a problem with such claims. Whether or not I can see these everyday things (and stars, for that matter) depends in part on the truth of following two claims:

(a) *Everything spatially related to me is present*: Nothing in the past or future is spatially related to me, although it might have been spatially related to me, and it might be spatially related to me in the future.

(b) *Everything that I can perceive or experience is present:* Nothing in the past or future can be perceived or experienced by me. However, it was perceived or experienced and will be perceived or experienced. Again, this means I cannot perceive or experience events such as the bombing of Hiroshima (to use an example from Houts 1980); such events are past.

I suspect (a) and (b) are either common-sense 'folk' claims or, at least, many think that they are common-sense claims. I suspect many will think it strange to hold that

- Contrary to (a), there are non-present things spatially related to me or my subjective activity now. This suggests that, as I sit here writing this in this café, the people who were sitting yesterday in the seats to my left (which are empty now) or will be sitting in the seats to my right tomorrow (which are empty now) are spatially related to me.
- Contrary to (b), the past or future can be perceived. Again, this would suggest that that I could perceive many things—e.g., those people who were or will be sitting in the seats to either side of me.

Yet although both (a) and (b) may be common sense or intuitive, and the contrary views sound strange, I want to argue that we should deny both (a) and (b). Furthermore, we can still preserve the idea that what is around us is present. However, this is a very weak idea of 'present'—one that neither threatens the intuitive claims about perceiving external things nor ameliorates the counter-intuitive claims of being spatially related, and of perceiving, things that are past.

6.2.1. Spatial Relations and the Past

Various theorists have asserted or assumed the claim that only present things are spatially related to present things.

Saint Augustine thought that past and future things, when they are past and future, could not be located anywhere. We do not observe them anywhere; we only observe present things. What we observe, of course, may very well be in the future and past as well. This is to exist beyond the present moment, which many persistent everyday things do. For example, when I meet you for the first time, you already have a history and so occupy the past in some way. The point here, however, is that I do not observe you *being* in the past; I only observe you *in the present*—i.e., (b).

One reason we may not observe something, however, is that it is hidden. That something is hidden suggests that it has some location (one which is unobservable). Thus one might argue perhaps that past and future things are hidden in some way. However, Augustine continues, even if they are hidden in some place, they are hidden there in the *present* and so are no longer future nor yet past (St. Augustine 1982). Thus although they might also be in the past or the future (just like observable things), such hidden things are only so hidden—i.e., spatially located where they are hidden, when they are *present*—i.e., (a) noted earlier.

Thus past or future things are not perceptible or spatially related to us *when* they are past or future. They can, of course, be perceptible or spatially related when they are present.

Houts invokes this thinking about spatial location in his 1980 paper on time lag (more of which I discuss further on). The city of Hiroshima is both in the present and in the past. Given '(a)' and '(b)', as a present city, it is spatially related to us. What happens in it now—e.g., people going to work every day, tourists visiting the memorial park—is spatially related to us. From my location here in Dublin, it is roughly 6,000 miles away.

Houts asks us next to consider the bombing of Hiroshima in 1945. Is anyone present—such as myself here in Dublin—spatially related to the 1945 bombing of Hiroshima? No, he states; this *bombing*, the event, is past, even if the city itself is also present. The bombing is just one moment in the past; we could pick any other moment in the past, and still Hiroshima at that moment would not be spatially related to us. The point is that a present thing, in the *present*, is spatially related to us; the present thing, in the *past*, is not spatially related to us.

We can more formally express this temporal constraint upon spatial relations in neo-Newtonianism terms. Neo-Newtonianism is a formulation of space-time in which one divides space-time into a series of simultaneous, spatially related events.

This concept of space-time has it that only simultaneous objects and events are spatially related. Objects and events at different times are not spatially related to each other. Thus we get this: If one event is past and another event is present, these events are not simultaneous. As such, given neo-Newtonianism, they are not spatially related to each other (e.g., Lange (2002), who also refers to it as *Galilean space-time*).

If we follow Augustine, Houts or neo-Newtonianism, then it seems that we are committed to (a), that we are not spatially related to past or future things.

That Augustine, Houts or neo-Newtonianism require it of us does not mean that we should hold that we are not spatially related to past or future things. Neo-Newtonianism is not the necessary or even the only concept of space-time. Further, it is not even the current accepted concept of space-time. That is the concept of space-time in relativistic physics. And relativistic physics suggests that we do not hold that only simultaneous events can be spatially related.

6.2.2. *Relativity and Spatial Relations*

In the previous chapter, I discussed the following thought experiment:

- Two spatially separated events, e_1 and e_2, are not simultaneous in one frame (frame A) and are simultaneous in another frame (frame B).

- Depending on the frame, these spatially related events either are or are not simultaneous.

If this is right, we get the following:

(1) The e_1 and e_2 are spatially related independent of what frame is defined.
(2) The e_1 and e_2 are simultaneous depending on what frame is defined.
(3) That these events are spatially related is *not* dependent on the frames the way simultaneity is.

If one to three are right, then it follows that the spatial relationship between e_1 and e_2 (which is frame-independent) is not defined or dependent on the simultaneity of e_1 and e_2 (which is frame-dependent).

This also gives the following:

- Even if one of these events is present, the other event, not necessarily being simultaneous, is not necessarily present.
- Yet both are spatially related to each other.
- As such, as with simultaneity, if one event is present, another event need not be present to be spatially related to it.

The final point here is this: Even if it turns out that relativity is false, unless it is false because it is *inconceivable*, then the conceivability of relativity theory show is that it is possible that things at different times are spatially related. It is possible to deny (a). In any case, it is currently thought that relativity is true; therefore, one should actually deny (a). Thus one should hold that present things can be and actually are spatially related to past things.

6.2.3. *Where Are Past Things?*

That past things are spatially related to present things does not mean that they are spatially related to present things the way present things are spatially related to each other. It is more complicated than that. The complication is important when we come to giving accounts of temporally and spatially extended perceptions; such perceptions, being extended in time, can have constituents that are both spatially related yet one constituent is past to the other. As such, this complication has implications later for how (i) we think about our spatial relationship to what we perceive and (ii) that relationship can be erroneous—e.g., the phenomenology can be different from how we really are spatially related.

I might say that Hiroshima is 6,000 miles from here (Dublin). That looks right for Dublin *now* and Hiroshima *now*. Yet what about the Hiroshima in 1945? What is happening in Hiroshima at that time is past and not present. From the last section, we can say that it is spatially related to us now. Yet does *what happens in Hiroshima in 1945* also lie 6,000 miles from here?

The answer to that depends on which frame we chose to measure the distances.

The cities are 6,000 miles apart if we assume a frame of reference in which Earth is at rest—i.e., Earth's own inertial reference frame. According to that frame, Dublin and Hiroshima do not change position in relation to each other. As such, there is no change in the distance between where Dublin or Hiroshima is now and where Dublin or Hiroshima is then. If they are 6,000 miles apart now, then, according to Earth, they are 6,000 miles apart in 1945, and one of the cities at one time is 6,000 miles from the other city at another time.

This is just like the example from the chapter on relativity concerning the mirrors and the observer on the platform. All of their spatial distances to each other remain the same through the moments when light hits one mirror and light hits the next. There is no relative motion between the mirrors and the observer on the platform during the light-reflecting events. So, for all three, the distances between light reflecting off one mirror and light reflecting off another mirror is the same at the time of each event (ignoring plate tectonics at least).

Yet we need not pick the frame of the platform to judge the distance between the mirrors and the observer. We can also pick the observer on the train moving through the station. There is relative motion between that observer and the mirrors. The distance between the observer and mirrors at one time and a later time has changed. In that frame, where one mirror is at one moment is not the same as where it is at the next. For example, if at t_1, the first mirror is at the front of the train, by t_2, it might be at the end of the train.

As with the mirrors and observer, we need not pick Earth's inertial frame to judge distances between Hiroshima and Dublin as they are at different times. Earth's frame is as arbitrary as any other frame. The distance between things according to Earth is as arbitrary as any other frame.

This is so even regarding events on its own surface. The distances between two places on earth depend on which frame one picks to measure them. If one does so from a fixed point on the sun, then the results are different to doing so from a fixed point on Earth. Yet both are equally correct measurements. It just depends on one's frame.

This may be hard to see with Earth. Earth seems to be at rest to those of us who are bound to it. It is tempting to assume that the frame of Earth has some special value with respect to points fixed upon it. It is certainly simpler to calculate distances over time between two points from a frame in which both points are treated as fixed. In that case, the difference in time makes no difference to the spatial distance. There is no change in distance over that time, and so we can ignore the time.

Yet simpler does not mean privileged and certainly does not mean absolute or real. Consider a different example: the frame of a car's wheel while the car is in motion. Say that a stone catches in the tread of the car's tire. Relative to the car's chassis, the axle through the wheel, and the road, the stone caught in the tire changes position from one moment to the next. As such, from one moment to the next, one can measure an appreciable non-zero distance between distinctly different locations of the stone. The stone is, for example, closer to the bonnet than the road at one time, then closer to the road than the bonnet at the next. The stone moves through this distance. There is a distance between the stone at one time and at another time.

However, the stone does not move through a distance according to the surface of the tire. It is embedded in the tire and at rest in it; it does not move through the tire as the tire moves. From one moment to the next, relative to the stone and the tire in which it is embedded, the stone does not move. Instead, the car, the ground, and everything else moving relative to wheel and stone are moving.

Yet this does not mean the following: The motion of the stone relative to the wheel is the *real* motion of the stone; the motion of the stone relative to the chassis or axle is an unreal or *false* motion. Nor does it mean that the *real* distance between the stone at one time and the stone at another is the real distance according to the stone, whereas the distance it moves through according to the chassis or the ground is *not* a real distance. The real distance is relative and no more the distance according to the wheel as it is according to the chassis. If there were absolute space, then perhaps one could hold that there is a real distance. Yet relativity denies such an absolute space.

Similarly, that a body on Earth is at rest or in motion relative to Earth does not mean that it is *really* at rest or in motion, against which we must judge other frames as false. It is really at rest or in motion, but only relative to Earth. Nor does it mean that its distance relative to Earth at one time and relative to Earth at another is the real distance. If we pick another frame—e.g., the centre of the sun—it is just as legitimate for measuring such distances.

Thus if we so wished, for the Dublin-Hiroshima distance, we could use the centre of the sun facing the galaxy of Andromeda to calculate where Hiroshima in 1945 is located relative to my writing this in Dublin now. As the constant rotation of Earth about the sun means that, relative to any such point on the sun, Earth's position is constantly changing. From one moment to the next, the exact same place on Earth—e.g., Dublin—moves hundreds to millions of miles apart, depending on what that point is and what Earth is doing. As such, the distance between where Dublin is at the time of the bombing of Hiroshima in 1945 and where Dublin is now can change dramatically. So can the locations of Hiroshima now in relation to that event and so can the distance between Dublin now and that event, or Hiroshima now and Dublin in 1945.

The answer, then, to the question of what is the distance between Dublin 2017 and Hiroshima 1945 is as dependent on an arbitrary frame as an answer about how fast something moves. There is only a frame-dependent answer to what the actual distance is between Dublin 2012 and Hiroshima in 1945.

Might this be a reason to deny that they are spatially related at all? Not at all. The problem of deciding on distance between events at different times does not promote rejection of spatial relations between events at different times; it presupposes them. The relationship between A and B is complicated. That it is complicated *denies* that there is *no* spatial relationship between A and B. (That *a* maths problem is complicated does not mean that there is no maths problem). This analysis of spatial relations over time goes as follows:

(1) There is a spatial relation between Dublin 2012 and Hiroshima 1945.
(2) However, one's frame of motion defines the precise spatial distance between these two points, at two different times.
(3) This is not as simple a situation as the spatial relations between things at the same time.

These discussions suggest that the idea that we are spatially related to things at other times is not, a priori, false, albeit how we are spatially related is complicated. However, it may still be strange for some, and thus worth resisting in any case. So are there any other reasons to deny spatial relations over time?

6.3. Concepts of Time and Spatial Relations

Here is an intuitive idea, and one I do not question in this book: I am not spatially related to what is only imaginary. By saying 'only imaginary', I do not mean imagining the Eiffel Tower (to use an example from McGinn 2004). I know I am spatially related to the Eiffel Tower. Yet imagination has nothing to do with that. I am spatially related to it because it is a physically real thing. What I mean by something 'only imaginary' is something like a dragon, a unicorn, or a moon base in 1999. None of these is real, and therefore—I assume here—none is spatially related to me. Similarly, I am not spatially related to merely hallucinated or illusory entities.

True, I can *imagine* being spatially related to these things. Yet this makes the spatial relation—which is only imagined—as intentional and unreal as to what I am related. It is still the case that I am not *really* spatially related to unicorns; I am only *imagining* being spatially related to unicorns.

Spatial relations are not intentional relations; if they really occur, they have real relata. As such, real spatial relations do not have unreal relata. We might hold the following principle:

> **The Reality of Spatial Relata:** If there is a real R and an unreal U, then it is not the case that R is spatially related to U.

Once we have this principle, there is one motivation to deny that we are spatially related to past or future things. Past and future things are not real. That is, presentism motivates our denial.

6.3.1. Presentism

With presentism, past events are unreal. There are no such events. We might say that we are several feet from a place where I once played as a child, but this does not mean that I am spatially related to my playing as a child. This would be as untrue as to say that imagined unicorns in my garden are several feet away from me, and so that I am spatially related to them.[2]

However, presentism works for this only because of its denial of the reality of other times than the unique present. With eternalism, other times are as real as such a present. So do we have a reason from eternalism to deny spatial relations over time?

6.3.2. Tense Theory

Perhaps tense theory promotes the denial of spatial relation to past things. However, if one accepts eternalism, then there seems to be no reason to enforce this.

Events and things which move through the A-series, going from the future to the present, from the past to the far past, etc., change in their temporal properties—e.g., in their positions in the A-series. This means one of two things for the other properties of said events. Either the temporal properties at least partly determine the other properties, or they do not. If they do partly determine them, then the change from one position to another makes a difference in the other properties of that event. If they do not partly determine them, there is no difference.

So when such events, standing in spatial relations, move into the past, do they change in their spatial relations to other events?

If events at one time are spatially related to events at another time, or if they are not so spatially related, their change in the A-series only makes them change in when they occur in the A-series (in their pastness, futurity, etc.). It does not otherwise change the events. Indeed, this is part of why McTaggart initially thought that B-theory descriptions of change were inadequate, and there is no sign that other A-theorists disagree with him on this.

To use McTaggart's example: A poker that is hot on the Tuesday and cold on the Wednesday always is, was, and will be hot on the Tuesday and cold on the Wednesday. This does not change. Similarly, the spatial relations do not change. If the poker end on the Tuesday is in the same place as it is on the Wednesday, this is, was, and always will be true. This does not change, either—at least, according to tense theorists.

6.3.3. Tenseless Theory

What about the position preferred in this book? Tenseless theory? Is there reason to hold that this position denies spatial relations between different times?

We already see reasons to have spatial relations given relativity, and relativity is a common reason to be a tenseless theorist. (Of course, one may put it aside and insist on a tenseless theory where only present things are spatially related. Yet it is not clear why one would, and certainly not why one would need to, do it).

To summarise, we can say this:

(1) The idea that we are only spatially related to present things and events may be intuitive. One is committed to it if one holds to presentism. As with imaginary or other unreal things, we do not stand in spatial relation to things and events at only non-present times—e.g., such as the past and future.
(2) However, given eternalism, one should allow that we are spatially related to, in tensed terms, past things and events (and perhaps future things and events), or, in tenseless terms, things and events at other times. In particular, if one holds eternalism due to relativistic physics, one must hold that we are spatially related to things and events at more than one time (whatever the moment that might be considered present).

This brings us on to the other question. I may be spatially related to past things, but do I perceive such past things? If we want to perceive what we seem to perceive, we should say that we do.

6.4. Perceiving Present and External Things

Perhaps there are things at other times spatially related to us. However, one might think, it is unimportant on the matter of naïve realism and on the constitution or veridicality of experience.

One might think this based on the following reasoning:

(1) Everything I perceive seems present to me.
(2) This raises two possibilities:

  (a) If how things seem is also how things are, then, just like distant aliens, non-present things and events are not perceived by me.
  (b) If what seems present to me are these non-present things and events, then my experience is erroneous; as with real but distant aliens appearing to be outside my window, these non-present things and events appear to be present, which they are not.

(3) If I phenomenology motivates my position, the preferred answer is '2a', that I only perceive present things.
(4) Thus, given naïve realism, only external things that are present constitute my perceptual experience.
(5) Thus, given naïve realism, even if I am spatially related to past things, they are not the part of what I perceive which constitutes my perceptual experience.

In response to this, we need to think about time. As discussed, given relativistic physics, it is unlikely that perceptual experience is constituted by spatially separated and simultaneous entities. Such simultaneity is arbitrary and, thus, not the sort of relation that plays a role in constitution. As such, it is unlikely that perceptual experience is constituted by entities which are not, according to some frame of reference, in the past (or indeed future) of other parts.[3] However, let us leave that aside. There is another reason to deny that we perceptually experience past things, despite their apparent presentness. There is a time lag between what we appear to perceive and anything happening within us that constitutes or appears simultaneous with such perception.

6.4.1. *The Presentness of What We Perceive*

As a first claim about appearances, let us accept the following:

(i) What we perceive appears to be only present.[4] It does not appear to be past or future.

With 'present' understood a certain way, (i) renders the claim that we are spatially related to non-present things irrelevant to perceptual phenomenology. Whatever our spatial relation to what is not present, in perceiving, we seem only to be spatially related to present things and events.

However, combined with two other phenomenological features, the situation changes:

(ii) What we perceive seems to be real.
(iii) What we perceive seems to be external and so spatially related to us.

How we analyse these three phenomenological features of perception—presentness, reality, and externality—is important. It provides test cases in which we show how time is relevant to experience more generally, as well as unveiling different ways of thinking about experience in time. However, it also shows how bringing many apparent properties together can take a toll on any claims that some metaphysical position is consistent with phenomenology.

Do we have any reason to deny any of (i), (ii), and (iii)? Yes, we should deny presentness—or at least a 'presentness' that threatens naïve models of perceptual experience. If tenseless theory is true, there are reasons to question the phenomenology of presentness as involving real presentness (particularly, as already discussed, given relativity). Yet there is an independent reason to deny presentness because of (i) and (ii). One cannot give an account of the real structure of perceptual experience that corresponds to the conjunct of (i), (ii), and (iii).

6.4.2. *Different Kinds of Present*

One can respond to the challenge from the phenomenology of presentness in one of two ways.

First, one can question whether there is any such thing as an appearance of presentness. Perhaps there needs to be a real present, but there is need for there to be 'presentness' of what we experience?

Second, one can challenge the assumption about what it means for something to be 'present'. If the present is indexed, as tenseless theorists hold it to be, why can we not conceive of past things as present as well?

(1) The Need for a Fundamental Appearance of Presentness

There is another way of looking at presentness in what we perceive. Mellor, Callender, and Le Poidevin question the need for the presentness to be anything over and above what comes from our perceiving it. X is perceived and X is present are the same thing.

Describe everything that is perceptually apparent—e.g., colours, shape, location, motion, timbre, texture, smell, and weight. Is presentness on that list?

Recently, tense theorists argue that the onus is on tense theorists to show that it is—i.e., that there is a perceptual appearance of presentness over and above the perceptual appearance of other properties.

Tense theory needs to argue that perception requires or even includes the appearance of an extra property called 'presentness'. Once we list what is perceptually apparent that is *not* presentness, there is nothing more to the experience, and there is no need to explain appearances with presentness.

If such tenseless theorists are right, then there is nothing to the perceptual appearance of presentness other than the perceptual appearance of something. It is not an extra property that is apparent to us in some way, permeating all that we perceive. It is just all that we perceive. Alternatively, if there is such a property, its supporters need to give reason to think so. Callender writes,

> Is being present a phenomenal property? Right now I sense my computer humming, black and white colors, and pain in my lower back, but I don't perceive a stamp of presentness on any experience. I am not at all sure I have an experience of presentness—as opposed to simply at present having an experience. Moreover [. . .] objects look the same to us even if they are in the past. The experienced point of light from Jupiter is of an object an hour in the past. Other points of light in my visual field may look the same even if the source is roughly present or something a million years old. Since events look the same despite tremendous differences in age, the view that we're sensing the property of being present of objects/events seems problematic.
>
> (Callender 2008, 3)

That is, given the actual physical world, naïve realists have reason to want that we perceive past things.

(2) Presentness Can Have Any Duration

As discussed in the chapter on time, tenseless theorists tend to think of 'present', 'past', and 'future' as being relative or indexical terms. Tenseless times define the 'now' or the 'present'. Further, the length of the defining time determines the length of the present. The present is punctal as defined by an instantaneous or punctal moment; the present is durational as defined by a period with duration.

A form[5] of durational present that is importantly relevant to perceptual experience is the *specious present*. The specious present is so-called because it is held to be an apparent present that is not really present at all. It has a long history in psychological and metaphysical discussion (over 100 years), with a few related but different meanings. Currently, many theorists hold it to be the period of time that we experience when we perceive change or succession.[6]

Here is one way to pick out the specious present: Wave your hand in front of your face. As you see it wave—i.e., move—you perceive the motion. Just as by seeing the shape of an apple you also see the spatial extent that shape occupies, so in seeing the motion of your hand you also see the temporal extent—or duration—that motion occupies. Thus as you see an extent of space in the shape of the apple, so you see an extent of time in seeing its motion. This perceived temporal extent or *period* is the specious present.

Different theorists have described the concept of the specious present variously and confusingly as incoherent and obvious. Interestingly, many of the claims about its incoherence seem dependent on tense theory. They rest primarily on the idea that only one moment can be present—i.e., the present moment is unique, even when it seems to be discussed without reference to tense—e.g., Sellars talks about the specious present being an incoherent mess of succession and simultaneity (Sellars 1968; for discussion, see Power 2013).

However, when considered from tenseless theory, many of these problems vanish. The idea of there being a number of moments which are present, even as they appear to be several moments, is no more a problem for a tense theorist than holding the following: that several spatial locations are here, even as separate locations. A region or volume of space can be indexed as 'here', yet have several sub-regions within it. I leave my car *here*, yet my car occupies a region of space. Similarly, for the tenseless theorist, a period of time can be indexed as 'now', yet have several subperiods within it.

6.4.3. *The Duration of Perceiving and What We Perceive*

However, with the perception of distant things, the main issue is not the duration of what we perceive. It is the duration between what happens inside when we perceive and what we perceive. If we accept naïve realism, then in constitutional terms, the main issue is the duration between different parts of our perceptual experience.

If we seem to see something which is not simultaneous with our seeing of it—i.e., not simultaneous with the neural events related to it, or the subject's ideas and thoughts (etc.) in experiencing it—then this clashes with appearances.

This is unlike the specious present. The specious present is a description of how things seem. The challenge for advocates of the specious present is to phenomenologically justify it and coherently describe it. Tenseless theorists may be capable of doing this. However, with our perception of external things, what a claimant to the idea that what we perceive is past suffers is that this is *not* how things seem. What we perceive does not seem to be at another time to our perceiving it. Whether or not we can coherently describe appearances, we do not have a correspondence between appearances and reality. We have a clash.

The phenomenology is as follows: The time when I am seeing distant stars shine, or the coffee in my cup going cold, is the same time as when the distant stars shine or the coffee in my cup goes cold. My seeing them and their occurrence are at the same time. As far as I can tell by having the experience, all of it constituents are simultaneous.

This makes perceptual experience unlike, for example, memory, where if I experience anything, it seems to be past—i.e., at a different time to me. If I were to hold, as with naïve realism about perception, naïve realism about memory experience, then the past event is a constituent of my experience. In that case, I might also hold that something at a different time to my remembering seems to be a constituent of the memory experience. Yet that is memory and this is perception: something at a different time does not seem to be a constituent of my perception or, indeed, my perceptual experience.

Thus, in perceiving external things, there seems to be no duration between something distant from me and my perceiving it. As such, something distant from me and my perceiving it seem to be simultaneous. This is the challenge of presentness transformed into a tenseless claim about simultaneity. Even if relativity relativizes simultaneity, there is still a clash with how things seem.

Therefore, we might think that, although some past things may be spatially related to us, and even may be *present* in tense theory, the duration between them and our perceiving runs counter to the phenomenology. Whereas, in both tense and tenseless theory, what is simultaneous with our perceiving of it does not run counter to the phenomenology.

This is why discussions about past things or things at other times being spatially related to us seem irrelevant to the phenomenological claim. It is because (a) we seem to perceive only things spatially related to us that are strictly simultaneous with our perceiving and so (b) as present as we the subjects of perception are ourselves.

Even so, we may want this because of one important empirical fact. This is the time lag in perception. To see its relevance, we need to talk about the standard physical account of perception. In so doing, we again introduce an account that is important background for later chapters.

6.5. The Process in Perceiving External Things

Light has a finite and frame-invariant speed (roughly about 3×10^8 metres/second). I discuss this in the chapter on relativity, including how its frame invariance has consequences for simultaneity. However, along with the invariance, there is also its *finiteness*. This does not have consequences for simultaneity generally. Yet it does have consequences for the simultaneity between subjective activity and the perceptually apparent.

This is unpacked in a rough model of the process of perception. Here, combined with one's metaphysics of time, this model raises problems with at least one of the three claims about what we perceive.

We can divide the process of perceiving something—e.g., seeing a distant star—into two main sequences of events. I call these sequences the *objective activity* (because the sequence is independent of the subject) and the *subjective activity* (because the sequence is dependent on the subject).

Each sequence has a number of sub-stages. Under the heading of each of these sequences, we can describe the perceptual process in terms of these stages. Let us take the example of seeing a distant star—a process depicted by Figure 6.1.

6.5.1. *Objective Activity*

First, there is the *objective activity* of the perceptual experience:

(1) At a particular time t_1, a star (X) at location (S_x) *shines* (H).[7] Therefore, the star shines at S_x (HXS_x). This shining is the emission of a group of photons from X that reach the subject (P). One can also describe it as photons that include P leaving the surface of X.

(2) P travels through space and various physical media (e.g., clouds of dust). Travelling, the path of P (M (P)) is affected by various things—e.g., gravity or the media it passes through. These effects alter its path. The waves of light containing P bend, refract, reflect, and, in some cases, are even absorbed. (We ignore the light that is absorbed; it never reaches the eye and so has no part of the overall perceptual process.)

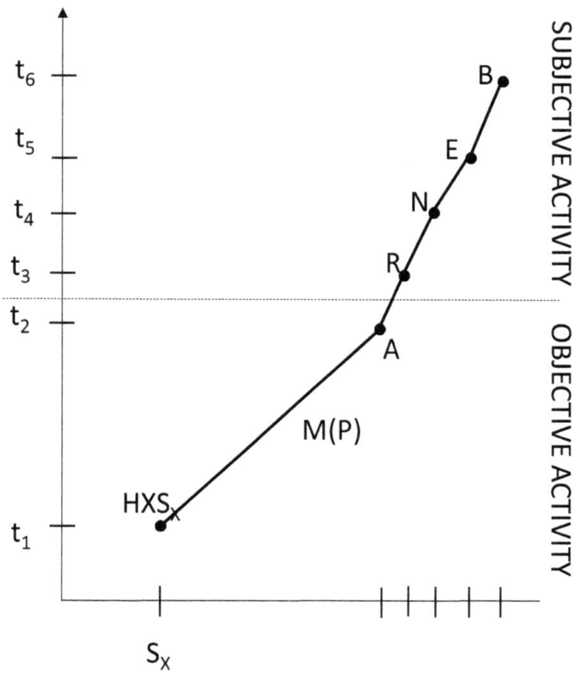

Figure 6.1 Subjective and objective activity in the perceptual process

(3) After some time, P reaches the retina of a subject S's eye (A). The particular time when it reaches S's retina is t_2.

Everything after t_2 happens in some way within S, the subject, and is subjective activity. Thus subjective activity begins after t_2.

6.5.2. *Subjective Activity*

Second (and last), there is the subjective activity of the perceptual experience:

(4) Reaching the retina, P stimulates the retinal cells, causing retinal activity R in S. The time when R begins is t_3.
(5) In turn, R causes further neural activity N in S. The time when N begins is t_4.
(6) In some cases, N either causes or is identical to S's experience (E). The time when this experience occurs is t_5.
(7) Because of E, S reports or exhibits other behaviour (B). The time when B occurs is (t_6).
(8) Most subjects report E as a case of *seeing the star* X *shining far away* ($E(HXS_x)$)—i.e., that they are experiencing the shining of the light in (1).

In the *reported* visual experience, the star's shining X (t_1) is at *the same time as* the visual experience of X (t_5). It may also be at the same time as any other subjective activity associated with the experience—e.g., unconscious neural activity that constitutes parts of the experience; it may be the same as other experiences that appear simultaneous with seeing the star (say, while gazing up, one is uncomfortably aware of a toothache). In addition, one might include as simultaneous with seeing the star the reporting of seeing it.

Once one is aware that it occurs, one might also include the light hitting the retina at t_2, or retinal activity such as R at t_3 or neural activity such as N at t_4.

One reason one may do this is that these events in the process of perception may appear simultaneous with each other and the star shining—i.e., it appears as if $t_1 = t_2 = t_3 = t_4 = t_5 = t_6$.

That is, it seems as if t_1 and t_6 (and the intermediary times (t_2 to t_5)) are all the same time. One might also say that these events are all *simultaneous*. Furthermore, it also seems that these events spread across several locations in space. For example, the shining star (HX) is far away from the retinal activity (R). When we examine them, R is not in the same place as the neural activity (N).

However, the finite speed of light means there is a time lag between each of these stages of the objective activity and the subjective activity;

they are stages in the light's passage from the star. This gives us one of four options. Each has different consequences for the relationship between appearances and reality:

(a) These stages are spatially related, just as they appear to be. However, they are at different times; this is not how they appear to be.
(b) These stages are simultaneous, just as they appear to be. However, they are not spatially related; this is not how they appear to be.
(c) These stages are simultaneous and spatially related, just as they appear to be. Empirical assertions that there is a lag in the process are in some way false. For example, the speed of light is infinite; there is no lag between retinal activity and any neural activity caused by that retinal activity.
(d) These stages are simultaneous, spatially related, just as they appear to be, and the empirical assertions about lags (such as about the speed of light) are true. However, a combination of psychophysics, biology, and neuroscience is false. The process of seeing does not include the stages described earlier—that is, the path of the light from the distant source, light stimulating the retina, or retinal activity stimulating neural activity. None of this is in the process of seeing shining distant stars.

Instead, there is some other process in seeing distant stars shining. Whatever it is, it means that what we see happening at distant stars is at the same time as the relevant subjective activity in our seeing.

These four options may be counter-intuitive or otherwise undesirable. However, they are exhaustive. One of them must be true. They show that we need to think about the appearance of simultaneity and its relationship to what else we seem to perceive.

6.5.3. *Presentism and Eternalism*

Let us take (a): What we see, and all the stages in between, are spatially related, but not simultaneous.

According to eternalism, something at one time is as real as something at any other time. Even if the present is unique (which tense theorists assert and tenseless theorists deny), it does not matter to an event or thing's reality if it is present or not. As such, if eternalism is true, the fact of a duration between distal and proximal stimuli, and the duration between these and neural activity inside the experiencing subject, does not alter the reality of any of these events.

According to presentism, however, only what is present is real; nothing at any other time is real. It very much matters to the reality of something whether or not it is present. If presentism is true, the fact of a duration between distal and proximal stimuli, and the duration between these and neural activity inside the experiencing subject, *does* alter the reality of some of these events.

Take the event that is the light emitted from the star. When the neural activity occurs, the emission of the light is past (more than four years past, since the nearest star is more than four light years away). Whatever we say about seeing the event, the time of seeing is connected to proximal stimulus; our seeing it extends at least over the neural activity and/or proximal stimulus (such as retinal effects).

However, given presentism, the light leaving the star is not a real event in perceptual experience; it is earlier than both neural activity and the proximal stimulus, events that are part of the experience. This leaves us with one of two alternatives. Either (a) we perceive this distant event, and so perceive something unreal, or (b) we perceive something real and we do not perceive this distant event.

Note here that this applies to *any* event removed from us in space. It can be the light reflecting off the moon, emitted from a street lamp, or a sound echoing from a cave wall. If you hear a sound from something several feet away, the sound may seem to be located there, to be a property of something there (e.g., O'Callaghan 2011). Yet if there is a causal link (with a non-zero duration) between that event and the resulting neural activity, and the neural activity is present and real, then given presentism, the distant event, not being present, is not real.

Butterfield argues that, for near enough events, there is no practical problem with this duration between the distant event and resulting neural activity. The duration is so short that, by the time the light reaches our eyes, nothing significant is likely to have changed. For example, a bouncing beach ball will not have moved through any appreciable difference in the time it takes light to leave its surface and reach our eyes. As such, for most events that we seem to perceive, there is no significant difference between what happens to stimulate neural activity and what is happening simultaneous with the neural activity (Butterfield 1984).[8]

However, even if there is no *appreciable* difference in events from one moment to the next, there is still a difference: there is still a duration. Causation over space takes time; this just is why there is a maximum causal 'speed limit'; no physical causal relations can cross between two points faster than the speed of light.[9]

As such, since presentism denies the reality of any moment except one, the present moment, for stages leading from distant stars to seeing stars, or bouncing beach balls to seeing beach balls, presentism denies the reality of all but one of those stages.

This means that, given presentism, we seem to have two options:

(1) What we perceive is not real. However, it is at a spatial distance.
(2) What we perceive is not at a distance from us. However, it can be real.

Option (1) runs counter to the idea that unreal things are not spatially related to us. One could still insist on it. However, it puts presentism at a disadvantage to eternalism. How it puts it at a disadvantage is evident in what (1) shares with (2). Both (1) and (2) run counter to perceptual appearances. With (1), given presentism, it is that what we perceive is unreal (being past). With (2), it is that what we perceive is not at a distance from us. Neither of these are how things seem, whichever we choose. Presentism commits us to one or the other.

It should be clear, that given time lag, presentism raises problems for the reality of what we perceive. Eternalism has some of these problems, as well, but to a much lesser extent. Compared to presentism, eternalism fares much better. There is less of a discrepancy between appearances and reality. The discrepancy for eternalism is this: It seems to us as if these distant objects are simultaneous with us; the star is shining now, even as I stand here looking at it. It seems to be simultaneous with my thinking about it, photographing it, and so on. However, it is not. It is happening at a distant point in the past or earlier from my thinking about it or photographing it.

However, with any position on time, this is the case here. It falls out of there being a time lag. Yet, with eternalism, there is this crucial difference: What appears to be distant and real is both *distant* and *real*. It is just that it happens at another time.

Still, even with eternalism, the aim of this work raises issues with what we seem to perceive and what we do perceive. I am not just trying to match the phenomenology of what one perceives. I am also trying to get at the constitution of perceptual experience itself. If what I perceive is such a constituent, then, to match the phenomenology of distant things, something at a different time to subjective activity is a constituent of my perceptual experience.

6.5.4. Tense and Tensed Theory

The tensed debate raises a different issue, even given eternalism. According to tense theory, there is only one present moment. So,

(1) If the subjective activity in me associated with my seeing (e.g., neural activity, thoughts about the seen thing) is present, then what I seem to see is not present.
(2) If what I seem to see seems present, then (one might hold that) this is an illusion. It is an illusion of *when* what I am seeing is located in time. I see it as being in the present, but it is in the past.
(3) Presentness and pastness, for tense theorists, are fundamental temporal properties of events, the very change of which gives us the fundamental flow of time.
(4) As such, this is an illusion of fundamental *temporal* properties of events—tenses.

(5) If what we perceive is distant and real, then, given both time lag and tensed theory, there is still a temporal illusion of tenses.

I return to this idea of being under an illusion of tense, and thus temporal illusion, in a later chapter. For now, let us consider tenseless theory. This returns us to the talk about the simultaneity of what seems spatially related to us.

With tenseless theory,

(1) If the subjective activity in me associated with my seeing (e.g., neural activity, thoughts about the seen thing) is at one time, then what I seem to see is not at that time.
(2) I mistakenly perceive the event as happening at the same time as—i.e., simultaneously with—the subjective activity in me associated with my seeing (e.g., neural activity, thoughts about the seen thing).[15]

This is the clash already discussed given time lag itself. It is because what is mistaken here is this: I fail to perceive the duration between the distant event and my own activity. However, nothing else is mistaken; if the event seems present, it is present. It just depends on the moment or period you pick to index it as present. Indeed, given tenseless theory, the perceived event is also past and future, as is the subjective activity; these are all relative terms. It just depends on the time you pick.

Again, the discrepancy between appearance and reality is that there is no duration between the event and seeing it. But how the event otherwise seems—it seems present, real, vivid, causal, active, or anything else you might think is part of the phenomenology of perception—this is, given tenseless eternalism, as right as if it were simultaneous with your seeing it. It *is* present, real, vivid, causal, and active. It is just not at the time when it seems to be your time as you see it and think about it.

Note the equivalence: It is not that the event in the past or earlier time is any of these properties, for example, real or causal. From some reason, one might hold that nothing is real or nothing is causal. The point here is that, if present or things simultaneous with subjective activity are real or causal, then so are past or things non-simultaneous with subjective activity. If they are identical in their non-temporal properties, then there is no more ontologically significant difference between these events than their tenseless temporal location.

6.6. Conclusion

To summarise the claims defended here about perceptual experience. Prior to this chapter, we have the following:

- *Perceptual experience is temporally embedded*. Both the subject and what the subject experiences appear to be in time.

- *Perceptual structure extends through external space.* To the perceiver, it seems to be of real external things. It seems to extend through space outside the subject to those real external things.

We go further than the phenomenology when we also consider claims about time: When we look at these claims given the fact of time lag, we use these appearances to justify holding that, as well as being embedded in time, perception is *extended* in time. So, from this chapter:

- Present things are spatially related to things at other times.
- Spatial relations between things at different times are not necessarily the same as spatially relations between things at the same time.
- Perception is extended in time because, to the perceiver, it seems to extend to real external things, which are at another time to the subjective activity in perception.

As such, we can say this: *To satisfy appearances, we should be eternalists, which allows real things at other times.*

These discussions show that being a presentist in a world with a time lag (never mind relativistic physics) can be problematic in satisfying the phenomenology of perception. Eternalism seems far less problematic. Thus it might seem as if eternalism is the better option here. As is probably obvious, this is my view.

Further, the ideas here, assuming eternalism, can be generalised to many cases of experience. Prior to consideration under such a general eternalist way of thinking, some of these situations are interpreted as illusions or hallucinations. Yet placed in an eternalist context, there is scope to interpret them instead as anosognosic.

We can see this by considering other kinds of experienced space. In the next chapter, we turn to a common erroneous experience of space: the 'distorted' perceptual experience of space. With the consequences of eternalism defended here, our explanations of such experience can be very different to common accounts in a way that further supports naïve realism.

Notes

1. Phenomenologically, I see this without needing any comprehension, concept, or understanding of what I see. Of course, I might comprehend, understand, or conceptualise what I see along with seeing it, but this is something other than what is involved in my seeing it.
2. One might respond by saying that, unlike other unreal things, past things, while unreal, can be spatially related to present real things. I am back in the house where I grew up. My growing up is past, but is also in the same place that I am now. That implies a spatial relation between a present and past thing, despite one being unreal.

One motivation for this is that there are problems with matching the phenomenology given one denies spatial relations to things that are, in fact, past.

However, I take it that such a response is either special pleading or a confusion. It is special pleading if one is simply asserting that time is special in that way. It is a confusion if one thinks that this is not analogous to spatial relations with other unreal things. Other than the species of unreality – past unreality and whatever unicorns are – the opposition can ask a question that does not have an obvious answer: What is the difference between the intentional 'spatial relation' between (a) an imagined unicorn and myself and (b) the events of my childhood and myself?

3. As a reminder: this is about the temporal relationships between the constituents, not between the experience, its constituents and non-constituents, e.g., such as efficient causes of the whole experience (and so precede it in its entirety).
4. Different philosophers come to very different conclusions about the relationship between experience and time because of such a claim. Mellor holds that, because our experience is like this, we cannot tell the tense of events from (perceptually) experiencing them (Mellor 1998). Droege claims that our consciousness represents its contents as happening now (Droege 2009). I take it that my discussion here relates to their positions. However, I do not discuss those positions specifically here.
5. A durational present can be any period of time that is present. For example, it could be the period containing the entire lifetime of the planet Earth.
6. Furthermore, Dainton considers this the period we experience even without perceiving change, such when we hear a droning sound: "Even though the tone does not vary in pitch, tone or timbre, we directly experience it continuing on" (Dainton 2000, 114). For this discussion, it is not necessary to consider this (in my opinion, likely) experience as well.
7. The light as it leaves the stars is an example of what many perceptual theorists call *distal stimuli* while the stimulation of the retina is an example of *proximal stimuli*. There are analogous scenarios for hearing, and the other senses: a process of distal stimulus, e.g., a door slamming shut, followed by proximal stimulus, e.g., sound waves affecting the eardrum. Somehow, this leads somehow to sounds (or tastes, feelings, smells etc.).
8. This conflicts with a recent hypothesis about why illusions happen. This hypothesis is that illusions are a result of compensation for durational gaps (Changizi et al. 2008; Changizi 2010).
9. Unless that there are exotic entities that move faster than the speed of light – *tachyons*. Tachyons are only speculative physical entities that, in relativistic physics, have strange properties such as a passage backwards in time. I do not consider them here.

7 The Perceptual Experience of Depth

If the last few chapters are correct, then, in order to perceive the external things that we appear to perceive, we should hold the position of eternalism. We should hold that we are spatially related to—and perceiving—past things. In addition, so that presentness in the phenomenology of perception is unproblematic, we should hold the position tenseless theory.

In the next few chapters, I use these assumptions of eternalism and tenseless theory to consider the following further perceptually apparent features of external things:

(1) The phenomenology of the visual depth of external things.
(2) The actual depth of perceptible external things.
(3) The relationship between (1) and (2), setting up a discussion on perceptual error.

7.1. Seeing Out There What's Not Out There

I am sitting in a café in Dublin with a view out the window. There is an empty lot on the other side of the street. Above this block, there is a grey sky. If I (and I am not sure how else to put this) look at the air above the lot *in a certain way*, it seems as if there is a slightly illuminated, translucent sphere hanging there. It hangs there steadily, even though a slight wind is clearly moving the bare branches of the trees beyond.

I know that there is not some strange device hanging up there above the lot. What it is of course is the reflection in the window of a lamp in this café. The dull day outside coupled with the fact the light is a source of illumination means that the light is visible even though I can also see what it outside the window.

However, although I know this, it seems remarkably easy for it to seem to me that the light is hanging out there above the lot. That is how it seems—it seems to be out there. Yet because I know about mirrors and reflection, etc., I know or judge that it is not out there.

There is nothing visible out there above the lot. The light is in fact behind or above me, and the air above the lot is transparent. Some of the light coming from behind me towards the window is being reflected back at me to hit my retina. If it were not reflected, I could not see the light, since my eyes are not pointing back at the direction of the light source (at least as I define it here).

However, in addition, I am not seeing the lamp *as if* it were coming from behind me. The lamp seems to be out there in front of me across the street. I might know that it is behind me (which, of course, I do), but that doesn't change how it seems.

From this, we can say,

(1) Something can appear to be a certain depth—or distance and direction—from me. The phenomenology can be of something, such as the lamp, as being a certain distance and direction from me.
(2) Yet compatible with that appearance, I can believe or judge that the thing is actually a different distance and direction from me.
(3) In contrast to both belief and appearance, there is the *actual* distance and direction of what I am looking at. I can believe it is in one direction, and it seems to be in another, and both belief and appearance can be wrong. In fact, the thing is elsewhere entirely.

> For example, I judge that the lamp is behind me, since I believe it to be reflected in the glass. However, the window is not directly in front of me; the pane is not at right angles to my line of sight. The pane is at a different angle to me such that it reflects things from a room to my left. From this, the lamp's actual distance and direction is some distance to my left; I believe the lamp is behind me; the lamp perceptually appears to be in front of me.

(4) This reference to both a distance and direction in how things are, appear and are judged to be is reference to coordinates defined along spatial *dimensions*. That is, when we define the depth and direction of what we perceive, we define it in relation to a coordinate system.
(5) One way to define these coordinates is Cartesian, in terms of three dimensions of width, height, and depth (this is not the same meaning of 'depth' as 'distance and direction'). The direction to my visual left is along the dimension of width; the direction above and below is along the dimension of height; the direction of before and behind is along the dimension of depth. The distance is determined from its distance along width, height, and depth.
(6) Another way to define these coordinates is radial, in terms of degrees of arc of circle with something (such as something about the subject) at the centre of that circle. In that case, something may be at a certain depth by being at a certain distance along a radius of the circle and in a certain direction by the arc of that radius. For example, something is to my extreme left by being at zero degrees of arc, straight in front

> of me by being at ninety degrees of arc, to my extreme right by being at one hundred and eighty degrees of arc.

So the lamp appears to be in a certain direction and distance to me—i.e., a certain depth to me. What do I mean that it appears to be at that depth? Under what conditions is that appearance mistaken?

To answer this, we need to consider a number of issues about the visual phenomenology of such depth.

First, the lamp may visually appear to be at a certain depth straight in front of me; it may also visually appear to be a certain depth above or below me and a certain depth to my left or right. For most of this discussion, let us focus only on the apparent depth of what is straight in front of me, the depth of what I am looking towards. With respect to error, I think that depth is a significantly problematic aspect of what we perceptually experience.

Second, the visual phenomenology of depth seems only to depend on what is visible. Merely by having some kind of visual perceptual experience, we have an experience of depth. This tempts me to say that *visual depth* is the area that is visually straight in front of the perceiver. It includes any distance along the *line of sight*. This depth (and any distance of it) is intrinsic to visual experience. The visual depth seems to depend only on having the visual experience itself; the visual depth seems also to be compatible with other things (e.g., beliefs or other experiences).

For example, what is *before* or *in front of* me is not just what is before or *in front of* my body. I might have a non-visual sense that my body is pointing east, and yet I am *looking* south. In this case, I would say that what seems visually before me is what is south, even though what seems before me bodily is in the east.

Similarly, looking at my feet, I might proprioceptively feel as if my feet are below me, but they visually seem to be straight in front of me; they are visually at some depth to me. Another example, walking through some woods, I look off to the left of my body while hearing a sound in front of my body. The sound might seem to be in front of me while what I see is off to the side of me; yet, even then, what is see is visually in front of me.

That is, vision experience itself appears to have its own direction and orientation, independent of my body and other kinds of experience. This is a visual appearance of depth independent of other kinds of depth. This is what it means for vision to have an intrinsic[1] depth.

One might question the claim that visual depth is actually *intrinsic* to vision. One might question the distinction between this 'line of sight' depth, visual width, and visual height. One might wonder if the distinction between them is actually intrinsic to vision—i.e., coming from vision itself—and not from a correspondence of what is visually apparent with something in the other senses—e.g., the orientation of our body, the vestibular system etc. One might question the position that the visual appearance of depth (and width and height) is from a single coordinate system or a single position in space. If it is not, each coordinate system or position, for anything seen, would give a different set of depths, widths, and heights. What would be at a depth according to one position could be at a height or width according to another.

However, for now, this is a phenomenological account of visual depth. In that case, these questions can either be ignored (if they concern actual visual depth) or met (if they concern appearances). There visually appears to be a distinction between depth, width, and height. There seem to be directions to what we see which are not defined by something other than vision itself.

At least, if there *is* such a visual phenomenology of depth, this is all very well. I have treated it as if there such a phenomenology. Yet is that the right way to think about the relationship between depth and what is visually apparent?

7.2. Apparent Depth

There is a great deal of literature on the relationship between judgements of depth and other properties of perceived things. No justice can be done to it. All I wish to do here is to clarify whether or not we are talking about the appearance of depth or of something else.

Despite my desiderata of saving the appearances and allowing for naïve realism, there are problems with saving the visual appearance of depth, where, again, depth refers to the space of what is *straight in front of me*.

First, the appearance of many kinds of properties that are not depth seem to be used for judgements about depth. This undermines the claim that depth itself must be apparent to us.

Second, even if we have a visual appearance of depth, it is quite unreliable. Many illusions seem to be based in some way on errors around depth. Furthermore, it looks as if these errors must be illusory. They cannot be easily reduced to anosognosia of depth.

The non-depth properties that seem sufficient for depth include:

(a) *Motion*: When we seem to move in relation to x, how x seems to move or how we seem to move in relation to x (or a combination) gives a compelling appearance of depth—e.g., if I watch a single spot of light in a dark room and the light traces a circle through the dimensions of width and height, then it seems to move away from and towards me through the dimension of depth (e.g., Kölers 1972). The movement of the light is not through that depth between the light and myself, yet how the thing moves is sufficient for generating that appearance of depth.

(b) *Uniform variation in width and height*: In terms only of width and height, a combination of relative size, regular in its variation, can give a sense of depth. For example, if one sees pebbles on a beach, if the pebbles are uniform in shape but, in terms of width and height, regularly decrease in size as one rises from the bottom to the top (e.g., Gibson 1950).

As well as giving the sense of depth, this changes how large objects seem. Objects can seem farther away and larger relative to other objects than they would if they were actually at the same distance. What might actually be different sizes actually seems to be the same sizes, just at different depths. What might actually be the same size seems to be different sizes and at different depths.

The latter result of such effects on depth-judgements can be used to explain various illusions—e.g., the Müller-Lyer, the Ponzo illusion, Ames rooms—and is a factor in explanations of more complex illusions (e.g., as discussed by Changizi (Changizi et al. 2008); for comment, see Briscoe 2010; Changizi 2010).

However, these features of things are not due to the apparent depth of things. Such appearances are in the width and height of what we see. We get the same effects with just width and height, as evidenced by most pictures, photographs, and films.

(c) *Contextual properties*: Something can seem to be a certain depth because of what surrounds it. This is most notable with the moon: The moon seems to be hanging just beyond objects in the horizon, or to hang in the branches of a tree. In this context, objects seem to inherit the depth of the objects they accompany.

Yet, again, if the judgement of depth is based on association with other objects, it is not a judgement based on the appearance of the object's *own* depth. The object's depth is parasitic on the other's objects' depths. We do of course have the depth of the other objects.

(d) *Colour effects*—e.g., *saturation*: The relative saturation or colour of some things in relation to others give a sense of depth—e.g., it is common that the bluer something seems, such as a mountain on a haze, the more distant it seems.

Again, colour is not the same as depth; contrast is not the same as depth. Yet, again, they are a factor in our judgements of the depth of what we visually experience.

Do such examples show that there is no need for the visual appearance of depth itself? Are the visual appearances of these properties sufficient for anything we might call the appearance of depth? If so, then there is no such appearance. This is because none of these are depth.

However, there are reasons to think that there is a visual appearance of depth:

(a) The aforementioned non-depth properties are not apparent. What is apparent is a mixture of instances of similar properties (but not those particular properties) and visual depth.
(b) Stereopsis requires the visual appearance of depth.
(c) There is a phenomenology of empty space, including in the direction of depth (i.e., what is visually in front of us).

I will take each of these reasons in turn.

a) The Appearance of Non-Depth Properties

It may be tempting to say that these properties (including relations) are perceptually apparent. That is, the motion, uniform variation, context relations, colour, and clarity are the way things seem to us. In that case, it looks right to say from the phenomenology of properties which are not depth properties, we make a judgement about the depth of what we perceive.

However, phenomenologically, when we come across these properties, they do not appear to be cases in which there is no appearance of depth at all. Instead, such properties seem to be similar properties instantiated at some depth or other. In some cases, this might make them appear to be the properties that they are. But in other cases, their seeming to be at some depth makes them seem to be some other property.

For example, take Kölers example: In an otherwise unlit room, subjects are presented with a light moving through an elliptical path only along the directions of width and height (relative to their direction of sight). Despite the absence of any other visual cues (since the room is otherwise dark), the light seems to move in an elliptical (but different) path through width, height, *and* depth. This apparent path of motion fills all three dimensions (depth, width, height); the actual path (relative to the subject) fills two dimensions; the paths are not the same.

That is, the actual property—here, the particular way it moves (in an ellipse perpendicular to our line of vision)—is not the apparent property (as a different elliptical motion not fully perpendicular to our line of vision). The difference is that the apparent involves motion through the visual dimension of depth.

However, one might insist that this is just a judgement of depth from an appearance of some other property. It is not just that there is some other property, which gives the appearance of depth. We are just very fast about inferring depth from the appearances.

For this to work here—i.e., phenomenologically—one must be saying that this is also how it seems: it seems to the subject that the subject is making a very fast judgement.

Yet this is not how it seems: Whatever properties, objects, relations, etc., that I see and judge as being at some depth from me, they seem to be at that visual depth. They do not seem to be at *no* depth. Things appear to occupy some point in the dimension of depth—e.g., a point in front of me, as vividly as they appear to occupy some point in the dimensions of height, e.g., above me, and width, e.g., to my left.

b) Stereopsis

Stereopsis is the sense of depth brought about by what is seen through two eyes (hence the 'stereo' in the name). It is not the sense of depth we get from something derived from one eye—e.g., colour, depth cues such as the context of other objects, motion, etc.

To those who have it, it seems to be somewhat close to a sensation all by itself—e.g., Oliver Sacks describes himself as having a strong sense of stereopsis (Sacks 2010). For others (such as myself), it is weaker.

Unlike, say, the faded blueness of a distant mountain relative to something closer, the depth of what is seen in stereopsis does not seem describable in terms of other properties. If one includes a sincere phenomenological description of how thing seems, in the case of what is apparent, one includes the phenomenology of stereopsis. As such, for those who have it, it is something indescribable in terms of other apparent things and, so, arguably, as much a feature of phenomenology as other shapes or colour or pain (for definition, see, e.g., Julesz 1966; Gordon 2004; Sacks 2010).

The existence of stereopsis suggest that depth itself is apparent to us. Given two normally operating eyes, there is an easily generated appearance of depth. It comes from the differences between what each eye detects.

However, this only works where there is such a difference in what each eye detects. Where there is not, we cannot draw on stereopsis to claim that there is an appearance of depth. Mountains, planes, planets, etc., and all such large and distant examples cannot fit within a range in which a difference between the eyes can be detected. Therefore, one might wonder if their apparent depth comes only from the appearance of other properties that they have—e.g., their colour.

This makes some intuitive sense. When we see the moon or distant stars, their apparent distance is not at all obvious. When we see the moon, the size and distance of it as seen from earth can vary by context (this is commonly called the *moon illusion*, e.g., Heelan 1983).

Yet we need to keep in mind a difference between the difficulty with the appearance of a *specific* visual depth that the moon has from us and the appearance that the moon has *some* depth. The latter is what we must deny if we are to deny the phenomenology of depth.

c) Apparent Empty Space

Mac Cumhaill argues that we depth is apparent to us even in cases where stereopsis does not apply. We seem to see *through* space, even empty space, to distant things (Mac Cumhaill 2013). We seem to do this even in cases where the distant things lie far beyond a detectable difference between the eyes (and so stereopsis). Mountains seem to be at a distance to us, irrespective of how faded they appear.

If Mac Cumhaill is right, the appearance of seeing through empty space requires the appearance of depth. This appearance of empty space is of a space with no other properties (it is not, for example, coloured), surrounding and extending beyond things which are visible to us, and irreducible to those things' properties, including their shapes, colour, or even their own depth. If this is right, then the appearance of empty space can only be the appearance of depth, width, and height. For empty space that appears to be in front of us—i.e., along the dimension of depth—it cannot appear to be just width and height. As such, it must be an appearance of depth.

To summarise, we can say that there is an appearance of depth in any perceptual experience in which

(a) Things appear to have some colour, size, shape, etc., at some depth to us.
(b) There is stereopsis.
(c) There seems to be empty space *through* which we see.

7.3. Hidden Depth

There are reasons to think that, along with apparent and actual (or real) depth, there is another kind of depth in the space between what we perceive and ourselves. This is *hidden depth*—the actual depth to what is perceived that is not apparent. Such hidden depth is external and real. It can also be the entirety of the actual depth of what we perceive. However, such hidden depth need not the entirety of the actual depth. It is possible that not all of the actual depth is hidden. Some of the actual depth might be apparent and not hidden from us.[2]

However, if it is held that none of the actual depth is apparent to us—i.e., that apparent depth is not actual depth—then hidden depth does encompass the whole of the actual depth of what we perceive. This follows trivially from denying naïve realism about depth—i.e., that the appearance of depth is any kind of depth itself. For example, if we conceive of apparent depth as determined only by actual non-depth properties, then, even if what we perceive is at a depth to us, its apparent depth from us is not any of the actual depth between us and what we perceive. Apparent depth is merely apparent depth.

Given a desideratum of this book is to hold on to naïve realism, I want that apparent depth not to be merely apparent. Unfortunately, there are significant problems with holding that naïve view. The actual depth of what we perceive must be hidden from us.

As discussed earlier, one gets apparent depth just by there being non-depth properties. Illusory examples are easy to come by—e.g., GOIs such as the Müller-Lyer illusion and Ebbinghaus illusion, and forced perspective illusions such as in the work of Bernard Pras. The illusions come only of the non-depth properties, yet seem to depend on an appearance of depth. They provide reasons to think that the conditions under which apparent depth occurs are independent from any real depth of what is perceived.

If this is right, then this also seems right: At least in these situations, apparent depth is not real depth; if it is anything, then it is something *other* than depth. Thus there is some reason to say that apparent depth is an internal construct or the result of some kind of unconscious or automatic judgement.

Although I would prefer direct or naïve realism about any properties which can be apparent, the following is unavoidable: We cannot be naïve realist about depth.

Indeed, strictly, we cannot even hold other perceptual theories about depth. For example, we cannot be sense-datum theorists about such depth. Sense-datum theory holds that we perceive the properties of external things indirectly through perceiving the properties of internal things, sense data. However, the depth of what we perceive cannot be captured by this. By definition, nothing internal can have such depth properties. These depths are between the subject and something in the external world.[3]

Thus, through something other than the depth of what we perceive, we make our judgements about the depth of what we perceive. Still, even if this indirect realism is unavoidable, there is a way in which we can evaluate our perceptions with respect to actual depth. We can evaluate the *judgements* about depth that come from perceptual experience. These judgements come from the 'apparent depth'—a depth that is merely apparent and thus (whatever it might be) not depth at all. However, although our judgements come from such mere appearances, they can still be correct.

I can judge from the apparent depth (which is not actual depth) of what I see that it is a certain depth from me—e.g., I judge that a farmhouse is farther from me than the road. This judgement can be correct; the farmhouse can be farther from me than the road. Albeit, it is still the case here that, because the apparent depth is not real depth, the real depth of what I perceive is wholly hidden from me.

With this, we can also say the opposite: I judge that what I perceive is a certain depth from me, but does not seem a certain depth from me. So I can ask, what discrepancies can there be between the apparent and actual depth—i.e., what erroneous experiences of depth can be had? And how should we treat these errors?

Appearances still matter here. These judgements are not made in a void, and just accidentally correlated with the real depth or not. The judgements come from the appearances; we make them depending on how things seem. For example, by the stereoptical phenomenology of some apples on a table in front of me, I judge that the apples are a certain distance from me. I appeal only to how it seems to justify what I believe. By the way, if the moon seems to be caught in the trees beyond the house looming in front of me, and thus seems to be a certain depth from me, then I judge that the moon is a certain distance from me.

One can treat these judgements as judgements from appearances and judgements due to apparent depth; they are made directly from the phenomenology. Thus it is possible that the apparent depth—even though it is not any amount of actual depth—matches actual depth. We can say this: It seems to me that what I perceive is a particular depth from me, and what I perceive is (or is not) really that particular depth from me.

7.4. Merely Apparent Depth and the Time-Lag Argument

As discussed in the chapter on the time-lag argument, given eternalism, even with the view that what we perceive must be real (and actual), we are not forced to deny that we can perceive external things. Things at a different time to us, e.g., even several years ago, are as real as ourselves, and so external things at a different time can be as real as ourselves. This puts eternalists at an advantage over presentists, because presentists must deny that these external things are real (this does not deny all external things are real, of course; present things can be real).

However, now we may have a reason to argue that this does not matter. The appearance of depth is an appearance of things at different distances to us in the external world. That is, the appearance of depth is the appearance of varieties of external distances. If apparent depth is not real depth, then it cannot be inferred from the appearance of depth that there is real depth. In the absence of this inference, one might try neutralising the argument from time lag to a preference for eternalism to presentism.

To do this, assuming the subjective activity in perceptual experience is present, one can argue as follows:

(a) The evidence of illusions of depth shows that apparent depth is consistent with no real (and actual) depth.
(b) The apparent depth of what we perceive is consistent with what we perceive having no real depth.
(c) If what we perceive has no real depth, then what we perceive is not external.

(d) If what we perceive is not external, then there need not be time lags between what we perceive and the subjective activity in perception.
(e) If there need not be a time lag between what we perceive and the subjective activity in perception, then what we perceive can be real and (strictly) present.
(f) It is the case that what we perceive has apparent depth but no real depth.
(g) (From (c) and (f)), it is the case that what we perceive is not external.

Therefore, ((d), (e), and (g)). What we perceive can be real and (strictly) present.

This may be a valid argument, but it is not obviously a sound one. It may be that the appearance of some object O's particular depth is consistent with O not having *that* particular depth, or even O having no real depth. However, by itself, this does not mean that O, in fact, has no real depth. That is, assertion (f) is an assertion that is not justified by merely apparent depth.

An alternative argument is possible. We substitute (a) with (a') and (f) with (f'):

(a') The evidence of illusions of depth shows that apparent depth is consistent with no depth or any depth at all.
(f') It is the case that what we perceive is external and has real (and actual) depth D, has apparent depth D', and D is not D'.

This does not get us the conclusion that what we perceive is real and (strictly) present.

A reason for holding (a) over (a') is that one holds that the non-depth properties giving apparent depth are inconsistent with real depth—e.g., some colour is inconsistent with having any depth at all. However, this is not at all obvious: An object's possession of certain shapes, context, motion, colours, etc., is consistent with it having *some* depth. After all, as we naïvely describe objects with such colours, they do have such depth. For example, although they need not be, the figures in the Müller-Lyer illusions are typically equidistant to the observer—i.e., they are at the same visual depth to the observer.

Note that an object's real depth (D in (f')) is not apparent to us (it is hidden from us). However, the *lack* of depth (in (f)) is also not apparent to us. All that is apparent is the depth that is denied by asserting either (f) or (f'). Thus if we are otherwise motivated to the view that what we perceive is external, that it does not have a particular depth it seems to have, that does not mean that we must deny its externality.

However, how do we know that there is a discrepancy between the apparent depth and the real depth anyhow? What justifies us saying that there is such a discrepancy? To answer that, we need to look at what the real depth of external things might be and then look at how it varies from apparent depth.

7.5. Conclusion

From this chapter, I conclude the following. External events that we visually perceive *seem* to lie at some depth from us. However, it does not look as if we can be naïve about this apparent depth; appearances themselves are not sufficient justification for holding that what we perceptually experience is at a depth from us and so external to us.

One reason why someone might want to deny that events are external and at a depth from us is because such a denial somewhat weakens the time-lag argument against presentism. Yet if one is not a presentist, it is not obvious that one must follow them into denying externalism. That the apparent depth of what we perceptually experience is a *merely* apparent depth does not mean that what we perceptually experience has *no* depth. If there are good reasons to hold that we perceptually experience the external world, then we should hold on to the metaphysical belief that, whatever the reliability of appearances, we perceptually experience things in the external world.

Yet if we *do* hold on to a perceptual experience of the external world, then it does seem as if what we perceptually experience is at a real depth from us. If so, there is both an appearance and a real depth to what we perceptually experience. So how do these relate to one another? And how should we understand the discrepancy between them? Does the apparent depth correspond, perhaps, to a distortion of the real depth?

To answer these questions, I assume that we perceptually experience external things. As already discussed, such a view is most compatible with tenseless eternalism. In assuming tenseless eternalism, I can hold that what we perceptually experience is

(i) External
(ii) Real
(iii) Present
(iv) At a different time to the subjective activity of our perceptual experience of it

The most important claims are (i) and (iv). If we accept that what we perceptually experience is at another time (and, often, another place) to what ourselves and what happens to us (subjective activity), then (i) and (iv) complicate the account we give of the actual depth of what we perceptually experience.

Notes

1. One might wonder what it means for distance or depth to be intrinsic. It is this: In visual experience, multiple entities can be at some distance or depth whether

the entities are otherwise accompanied or alone. For example, I appear to be at some depth to the surface of water that I see or at a distance to the lines on a road; this is compatible with my standing in these relations while I and the surface or I and lines of the road are otherwise alone or accompanied. If I and the surface or the road cannot be otherwise alone, I take it is because it is not possible for me to stand alone in any relation to a surface or a road.

2. Perhaps we could call it *obvious* depth. An obvious thing is both an apparent and real thing. (This is similar to my use of 'phenomenal' and 'phenomenal parts' in my 2010: I took 'phenomenal' there to mean both apparent and real, but I no longer do so).
3. I do not understand a concept of depth that is not spatial depth, and I do not understand a concept of spatial depth that is not physical depth. As such, I do not understand the concept of a non-physical depth, such as any that sense data might possess. I take it that it does not mean apparent depth, for that need not be depth at all, and, in any case, this is an appearance of physical depth.

8 Distortions of Depth

The discrepancy between (judgements from) apparent depth and real depth is complex to define. Once we include how our subjective activity is related to the external events in time, the picture only gets more complicated. In that case (as already discussed in the chapter on spatial relations over time), how what we perceptually experience is related to our subjective activity seems open to further arbitration and convention. Depending on the frame of motion we pick to determine it, the discrepancy between apparent and actual depth can vary significantly.

Further, even if we only focus on how our subjective activity is related to external events *right now* (however that is defined), it is difficult to define the real depth between such external events and ourselves. In the absence of absolute space, in which distances and straightness may be defined, there is a degree of arbitrariness and convention which suggests that any 'real' depth of anything is itself relative, arbitrary, and conventional.

However, this does not eliminate error. It appears that there is an absolute space; we can use that space to define a singular and real depth. Yet, even if there is a real depth, any discrepancy from it is not obvious to us. It comes from our perceptual experience failing to detect something that is there. Still, it does allow a re-interpretation of the error as involving anosognosia rather than illusion or hallucination.

In any case, I assume that our judgements of the depth of what we perceive do not come only from the perceptual appearance of depth. Our judgements can still correspond to the real and hidden depth of what we perceive and thus be judged as true or false.

How can the hidden and real depth vary from (our judgements from) apparent depth?

To answer this, consider mirages.

8.1. Conflated Points of View

An object that looks straight in the air is half-immersed in water. As a result, it looks as if there is a bend or break in it where it crosses the border between the water and the air. This is due to the light reflecting from the stick as it travels through the two separate mediums: the air and water.

Thus the stick appears to have a shape that it does not have. This is a *distorted* image of the stick. There is a discrepancy between the appearance and reality of the stick's shape.

We can treat the appearance of a bent or broken stick as an example of a *mirage* (e.g., Antony 2011). A mirage is a visual perceptual experience where (a) there is a discrepancy between appearances and reality and (b) this discrepancy is not due to subjective activity in the perceptual experience but due to something in the objective activity. It is due to the path light takes to reach the eye.

For many philosophers of perception, mirages are uncontroversial cases of an object having a shape that it does not appear to have. Indeed, arguably, many objects in mirages (including the bent or broken stick) appear to have shapes that are more complex the shapes they actually have; in addition, they appear to occupy dimensions that are greater than the dimensions they actually occupy.

Thus one might hold that such mirages are exemplars of illusory experiences.

I do not question that it is some discrepancy between appearance and reality. However, I do not think it is an illusion. It is a case of anosognosia. This anosognosia is due to the complexity of the objective activity in perceptual experience.

Returning to the bent or broken stick, there is a difference between the stick's actual or real shape and the stick's apparent shape. The stick's apparent shape is a *distortion* of its real shape because of how the light passes through the water.

One way to unpack this distortion of shape is as a distortion in the relative location of its parts. The bending of the light gives an appearance of the parts of the stick as having spatial relations to other parts of the stick. These apparent spatial relations are not the actual spatial relations between these parts.

The apparent locations for the parts of the stick can be broken into apparent depth, width, and height of the stick. The difference that is typically given between appearances and reality is that the part of the stick in the water seems to be somewhere other than it is.

However, is that correct? Is the real depth, width, and height of each part of the stick different to the depth, width, and height that it seems to have? To say so is to say that each part has a *real depth, width, and height*. Thus we can ask, what is the real depth, width, and height of each part of the stick (and thus of the stick overall)?

To answer that question, we need to know what it is that we are seeing and where it is in both space and time. In answering it, we see the discrepancy is more complicated than it initially appears to be.

8.1.1. Unusual Points of View

What are the conditions that must be met for there to be a discrepancy between apparent and real depth in a mirage (such as the apparently bent stick in water)?

The following are not the conditions that must be met for this. It is not sufficient for this discrepancy that the apparent depth of the object or its parts diverge from the following:

(a) The object or its parts' location now.
(b) Where the object or its parts appear to be in more common situations.
(c) Where the object or its parts appear to be from a single point of view.

a) Not Where the Object or Its Parts Are Now

I assume tenseless theory and that there can be a spatial depth between things at different times. I assume at least the possibility that naïve realism is true. Thus I hold that things at different times (e.g., subjective activity and external events) can constitute perceptual experience.

The result of these assumptions is that, if we do perceptually experience something at a different time (to subjective activity), then the relevant depth is not the depth between us and where what we see is spatially *now*. If we ask what the depth of what we see is, we should ask about where it is in the past, at the point when it is the source of light. It may be in the same place since then. It may have moved since then. However, where it is in the same place or somewhere else now is irrelevant to where it is as a constituent of our perceptual experience. Where it is in the past determines its real depth to us.

I suspect that this is not an intuitive position to take on the depth of what we perceive. The intuitive position is that the depth of what we perceive should be its depth *now*. However, the alternative position I advance here is a simple result of taking all three, naïve realism, time lag, and tenseless eternalism, seriously. If the alternative position is correct, but one holds fast to the intuitive position, then one may make judgements about apparent depth's accuracy that are mistaken.

For example, one may hold that the depth of something now is not the same as its apparent depth. Thus one concludes that its apparent depth is in error. Yet its apparent depth may be the same as the depth it is at the time it is the source of the light leading to the perceptual experience. In that case, if we do experience it when it is at that time, the apparent depth is accurate.

b) Where the Object or Its Parts Appear to Be in More Common Situations

One difference for most people between seeing the shape in the bent stick phenomenon is that it is not merely through the air. It is also seen through the water. This is a difference in medium and arguably less common. As already discussed, the commonness or uncommonness of an appearance does not make it either veridical or erroneous. However, if one does not know what the difference is, it might tempt one to think that there is an error when there is no error.

For example, there is a mountain I pass every day and always see from one particular angle. One day, I happen to approach it from some other side without knowing it (I got lost in a fog). I approached it from an unfamiliar angle. As a result, it was unrecognisable: initially, it seemed to be a different mountain because it was a different shape. When I recognised it as the same mountain (through a landmark on it), I thought that I was seeing it in some distorted way.

Yet my experience was not a case of distortion or a discrepancy between appearance and reality. It was only that I saw it from an unfamiliar location in space. I thought it was a distorted perception because I was ignorant of the following fact: *I was seeing it from a different location.* By presupposing that I was in the usual position in relation to the mountain, I thought the mountain was in its usual spatial relation to me. Thus I judged that its uncommon appearance must be different to how it really is—because how it really is, is how it commonly appears to me.

There is no discrepancy between appearance and reality here. The mountain looked as it did because I was seeing it from a *different point of view*. I was seeing it from the other side. When seen from my actual position to it, this is how the mountain looks.

What this experience can result in is, through assuming we are in the usual situation, I come to an wrong understanding of what is wrong. I think my perceptual experience is in error. Yet, my thinking is in error - about my perception and the situation that I am in.

We may say something about the appearance of a bent or broken stick from light passing through different mediums. We might hold that this is only a less common instance of the process involved in seeing. In this case, the particular mixture of media through which the light passes here (this mix of air and water) is less familiar to us than another mixture of media (e.g., air). The relative rarity of this mixture means that our interpretation of it in terms of how things commonly seem can lead to us interpreting it in different ways. However, its *rarity* is not sufficient for it to be a distortion. Perhaps given the point of view from which we view the stick, it is right to say that the stick *is* bent.

Distortion requires that the stick *really is* bent or straight. Again, it may seem obvious that the stick is straight and unified, and so its apparent curvature or brokenness is mistaken. Yet there are cases in which a stick can veridically appear as it does in the bent stick phenomenon and yet be straight. This concerns the complexity of our perspective in perceptual experience.

8.1.2. 'Cubist' Vision: Complex Points of View

When viewed from the angle that we normally see it, sticks are straight, unbent, and unbroken. However, there is also this possibility: When the stick is seen from an unusual angle, then the stick is appears bent or broken. There is another way to describe it, analogous with the mountain example. The stick does not appear to be bent or broken; this suggests a division or property of the stick, which has no such division or property. Instead, it is this: The parts of the stick appear to be seen from *different angles* or *different points of view*. The division or mistaken property is in something else.

This opens up a different account of the discrepancy. There is a discrepancy because (a) we are not aware that we are seeing it from different angles; as such, (b) we mistakenly hold that we are seeing it from the same angle. If it appears to us as if we are seeing it from the same angle, then this is an erroneous experience due to a failure to recognise the complexity of the constituents of our experience. When generalised, and applied to an actual case discussed here, this is anosognosia.

The anosognosia comes from a simple fact: The direction in the visual appearance of depth is not determined from observations of the direction in which our eyes look. The eyes are not apparent to us in visual experience itself. As Wittgenstein puts it, "From nothing in the field of sight can it be concluded that it is seen from an eye" (Wittgenstein 1974, §5.633).

However, the eyes are relevant in thinking about the real visual depth of things. One might take them as the point from which we see, and in doing so, although not themselves apparent, affect the case that what is visually apparent is accurate or inaccurate.

Note that this is not a claim that we can sufficiently define the real visual depth by the eye. It is that, if you are to define it, you need to start with the eye, and not something else—e.g., the body overall, or the brain, or the head (or simply choose the direction arbitrarily). Visually apparent directions of any kind, including depth, can diverge from directions defined by the latter entities without being in error—e.g., what is in front of my body can be different to what is visually in front of me.

So let us consider now a fantastical example of seeing where the definition of depth is more complex than typical human perception.

Early Cubist paintings often represent seeing something from different points of view. For example, Picasso's *The Weeping Woman* (1937). This is a portrait of a woman crying that is partially in profile and partially face on.

Under normal circumstances, this combination of visual points of view is impossible. Our eyes cannot be orientated about a scene to give a similar visual appearance; I cannot move one of my eyes to see your face from the side while the other sees you head-on. (Nor can I move one of my eyes to see a mountain from one side while my other eye moves to see it from another side.) We just do not have the physiology available for our eyes to do that on their own.

However, imagine a creature with what I shall call *cubist vision*. This creature is able to see something from the variety of perspectives simulated by a cubist painting—i.e., from significantly different points of view. A creature with cubist vision can manoeuvre its eyes around so that it can see from different angles. Its eyes are somewhat set up similar to our hands. It is, perhaps, like the monster in the film *Pan's Labyrinth* (2006). It can even point each of its eyes to look at the other, such that it can see one of its eyes with the other (just as I can grasp one of my hands with my other).

Next, I want to turn to the actual complexity in normal human vision. This is vision involving two eyes, both facing roughly the same direction, but with some possible variation in how depth from each eye is defined or can appear to the subject. With human vision, the two visual perspectives can be fused, mixed, a combination of fusion and mixture, or a matter of rivalry and suppression between them. Something similar may happen with creatures with cubist vision.

A. *Fusion*: What I see through each eye fuses together. What we see is due to a process similar to the process for normally orientated eyes that leads to stereopsis. With two eyes, this is stereopsis. For creatures with more than two eyes, it might be better to call it *polyopsis*.
B. *Mixture*: What I see through each does not fuse together. This is similar to what we see in cubist paintings. We see through each eye slightly different things, and they each take up some of our visual field. What we see through each eye *mixes* with what we see through the other eye. We see this in *The Weeping Woman*: We see her face partly in profile and partly face on.
C. *Combination*: A combination of A and B. We might even be able to change easily from fusion to mixture and back. Many people can do this with their own vision. They can cross their eyes. What is a fusion becomes 'seeing double'.
D. *Rivalry and Suppression*: In addition, there could be rivalry between the stimuli—a flickering back-and-forth in perceptual experience between the stimuli from each eye. While both are working, there could also be only a perceptual experience from the stimulus of one eye and not the other eye. The creature only experiences the side of a face, and then the front of the face, and then the side, and so on. Alternatively, the creature only experiences the side of the face (but, perhaps, still reacts—blindsight-like—to stimulus from the other eye).[1]

In principle, one can extend these to creatures with more than just two eyes. For example, one might have such fusion, mixture, or rivalry with three eyes (e.g., a pituary mystic), five eyes (e.g., the last men of Olaf Stapledon's 1930 novel *Last Men and First Men*), eight eyes (e.g., a spider), multiple synthetic eyes (e.g., Wallace in Deneuve's 2017 film *Bladerunner 2049*), and countless eyes (e.g., the ram in Russell's 1980 film *Altered States*; Brooks' multiply-bodied minds (Brooks 1994).

I am not sure what such visual experience would be like to have in itself or in relation to one's other experiential and perceptual modes—e.g., how it affects one's interaction with the world or what sort of body one would need to do so reliably. However, one way we might generate a similar experience is through mirrors. Consider the example in the opening of the last chapter of seeing the lamp hanging over the lot. This is what one would see with a mixture of two eyes, with one looking back and one looking forward. Perhaps fairground mirrors are like seeing through significantly and differently arranged and shaped arrays of retinal cells.

The important question for seeing depth is this: For such imaginary creatures, is there a discrepancy between what appears to be there and what is really there? Is what they see a distorted appearance of the visible world?

Not if it appears to the creatures that they are seeing things from all of these points of view. If that is how it appears to them, then how it appears to them is also how it is.

Here is an analogy: Consider your body. You feel as if your hand is somewhere other than your feet. Say you stand barefoot on grass while running your hand against a plastered wall. The grass can feel soft against you while the plaster feels hard against you. It appears to you as if parts of you—your feet—are touching the grass while other parts—your fingers—are touching the plaster wall. You are aware that you have multiple 'points of view' here, and so there is no error.

Similarly, a creature with multiple and prehensile eyes would not be mistaken if what it sees appears to be from multiple and separately moving points of view.

8.1.3. Erroneous Points of View

Yet for creatures with cubist vision, there can be a discrepancy between how things are and how things appear. If it appears to them that they do not see the world from multiple points of view, then this appearance conflicts with reality; they *do* see the world from multiple points of view. The distortion is due to leaving something important out: the differences in the points of view from each of the subject's eyes. This difference is a difference in perspective or point of view.

The absence of differences in points of view raises two possible appearances with respect to the point of view:

(a) A unified point of view.
(b) A 'view from nowhere'—i.e., the absence of any perspective or point of view at all.

If such a creature has the non-appearance of a multitude of perspectives, then it is either because it has the appearance of a unified, shared perspective or of no perspective at all.

If such creatures are like us, what they see does not appear to be seen from nowhere. It appears to be seen from somewhere. Thus we can discount (b). This leaves open the possibility of (a). The locations from where the creature sees appears to be one location. The different points from which it views are conflated into one point of view.

This distortion is not a distortion with respect to what is visually apparent from each eye and what is visually before each eye. The distortion is in there appearing to be no difference between the locations of its eyes.

Let us apply this to a cubist version of seeing a stick in air. Imagine that I am someone with cubist vision looking at a stick (in air). My eyes are positioned in such a manner that one eye is above the stick and one eye is to the side of the stick. In that case, I look at the stick from above and from the side. If I see the stick through air, what do I see?

From the side, I see the stick's entire length but from above I only see its thickness. Through one eye, I see the stick's length occupying height or width; through the other eye, I see the stick's thickness or cross-section occupying height or width.

As discussed, the visual appearance may result from their relating to one another in a number of ways: Each may fuse, mix, conflict with, or rival or suppress the other. I cannot imagine a fusion like this.

However, if it is a mixture, the stick might look like radically different things: From above, it may look like a relatively small circular thing; from the side, it may look like a relatively long and very thin rectangular thing.

This is the case for those with cubist vision—i.e., creatures with eyes unusually distributed in space. The difference in their locations and directions mean that, although the fusion or mix of what is seen may be mistaken—e.g., a rectangular/circular mix—what is seen by each eye is not mistaken. That they have a unified perspective is mistaken.

What sort of erroneous experience is this? This is not an illusion or hallucination. It is a conflationary error. It is a case of anosognosia. It is anosognosia of the multiplicity and complexity of the creature's own points of view.

8.2. Distortions of Hidden Depth

One might wonder why all of this matters. I do not have cubist vision, and I do see the broken or bent stick. So what relevance does this merely hypothetical vision have for the visual depth according to the real non-cubist me? There are two significant differences between the fantastical and the real here: At least with respect to broken sticks (and other mirages), the following seems right. With respect to seeing the stick, I have a unified perspective. Differences between the positions of the eyes are irrelevant.

This irrelevance also extends to the arrangements of other seen things—i.e., the apparent width and height, the kinds of colours and shapes. There is no fusion or mix of different perspectives, as there might be with cubist vision or stereopsis of crossing one's eyes. What is visually apparent to both eyes through water and air is a broken stick.

Therefore, through both eyes, I see the stick from practically the same point of view; for both eyes, the visual appearance is the same. So how could the apparently broken or bent stick be a case of hidden difference?

If we consider that, as described here, the apparent shape is more complex than the real shape, then it appears to be particularly implausible that the 'bent stick' error is due to a hidden difference. A bent or broken stick has a more complex shape than a straight stick.

Bent or broken sticks need two dimensions; if the stick is vertical, the break or curvature is partially horizontal. Straight sticks need only one dimension; they can fully occupy the vertical:

- The shape of a bent stick has an angle between its sections absent in the shape of the straight stick.
- The shape of a broken stick has the break, whatever it might be—e.g., a gap in the dimension of the stick's length or a shift in location along the second dimension; the straight stick does not have this break.

Anosognosia is an erroneous experience involving only something hidden. How can anosognosia of something hidden explain a discrepancy between what is really a straight stick and its appearance as broken or bent? As stated, this requires that the way some O appears is due to something not apparent or hidden about O. Yet with respect to the shape, there is less to O than there appears to be.

This suggests a problem similar to re-interpreting hallucinations as anosognosia. What is hidden is less than what is apparent. Even if it only involves properties such as shapes, as it does with mirages, it still at best raises the trouble of negative properties (there is an appearance of some negative property, such as the non-occupancy of a second dimension).

8.2.1. Distortions of Direction

Must we deny that there is anosognosia with broken sticks and mirages? No. We can appeal to anosognosia. It is just that it is not merely the anosognosia of the stick's *shape*.

Here is what is missing for people with normal vision—such as you and me: It is something very similar to cubist vision. It is the variation of *depth* in the object in perceptual experience, where I mean by 'depth' here the distance and direction of it. The missing depth is in the path of light involved in the perceptual experience. It is more complex than what is perceptually apparent to us. This explains such things as the bent or broken stick and other mirages.

I referred in the last chapter to depth as the distance in front of the visually perceiving subject: The lamp reflected in the window *seems* to be at a certain visual depth in front of me—the same visual depth as the lot across the street. This is a *single direction*, a *single* depth, width, and height. It is a single line of sight. This makes sense only if the various origins for the depth all face the same way, giving the sameness of direction.

However we define a single line of sight (e.g., our eyes 'face the same way'), the paths of light leading to our eyes from external objects need not lie along such a single line of sight. They can lie along diverging paths of light at the end of light paths that diverge from one another.

Thus one possible way that appearances and reality diverge is that the various paths of light diverge but do not seem to diverge. All of them seem to correspond to a single line of sight—e.g., the depth straight in front of me. In that case, we are not aware of this divergence. We fail to be aware of that complexity; we are only aware of a simpler depth between external things and ourselves.

This kind of erroneous experience is anosognosia; we are anosognosic of the curvature of light in space. If we are anosognosic of the curvature, then, due only to the limits of our perception, where there is really a straight stick, we can have perceptual experiences of a bent stick.

However, note that a time-lagged relationship to the external thing determines its real depth to us. The speed of light is finite, and there is a non-zero duration between a perceptual experience's subjective activity and the apparently perceived distant object. As stated at the beginning of this section, when determining the paths of light, we consider the actual paths that lead from the source at another time. We do not consider any paths from the source now.

Thus when asking if something at an apparent depth to us is indeed at that apparent depth to us, we can ask if the apparent depth of something from us corresponds to its real depth from us. Is what seems to be straight in front of us, and what seems not straight in front of us, really straight in front of us?

In cases such as mirages, the intuitive answer to this question is *no*. As already stated, unlike cubist vision, the difference between the apparent and actual depth does not seem to be because of the location of one's eyes.

A typical example of a mirage process is this: Light passing upward from an object hits a temperature inversion in the upper atmosphere and is refracted and/or reflected back towards the earth. For a subject whose eyes intercept this light, they see an object as displaced in space (and due to the way the light bends back, the reflecting object seems to be upside down). It does not seem to be nearer (or only nearer) than it really is; if we see it along the dimension of depth, the process does not simply move what already lies along it closer to us. The refraction or reflection means that what is not straight in front of us appears to be in front of us, and what is straight in front of us appears to be not straight in front of us.

This is fine. It is anosognosia of the real depth and direction of what we perceive. That depth and direction can involve imperceptible refractions, reflections, and other hidden effects of light's transmission. It leads to a discrepancy between appearance and reality but it is not one in which there is more to the appearances than is really there.

8.2.2. Geometric-Optical Illusion

Could one interpret GOIs as anosognosia similar to the way I have with mirages? Unfortunately, I do not think we can. These have to be illusions, with all the resulting epistemological consequences of that. We cannot re-interpret the erroneous appearances in GOIs to the non-appearance or incomplete appearance of something else.

To show this, I draw on a common explanation for why these experiences happen. This theory comes from Gregory (e.g., Gregory 1987, 1998; for discussion, Gordon 2004), who develops it out of work by Helmholtz. (Also, Changizi (e.g., Changizi et al. 2008; Changizi 2010) further develops it in a way that brings time and presence into the account.)

Similar to mirages, in GOIs, the experiences of the figures are not merely of the two-dimensional figures themselves—e.g., of the complex of circles in the Ebbinghaus. They are of those figures located in relation to the subject in three-dimensional space. When we look at such figures, we see them as in front of us and at some distance from us—i.e., at some depth from us.

As has been discussed, the appearance of depth is unreliable and is due in part by the non-depth properties of what we see. This is also true of the figures in GOIs. There is an appearance of their depth, and this appearance is due in part to the non-depth properties of the figures.

Similar to the case of the apparently bent or broken stick, we can understand many of the GOIs as apparently relative spatial relations between the figures. For example, the Ebbinghaus illusion concerns relative circle sizes (they appear to be different sizes but are the same size), the Hering illusion the relative direction of two lines (they do not seem to be parallel but are parallel), and the Müller-Lyer the relative lengths of two lines (they do not seem to be of equal length).

Despite these figures being two-dimensional, they are still in three dimension. According to Gregory, the erroneous appearances are the result of something built into the perceptual process, an assumption of how such appearances occur in three dimensions. The basis of this assumption is what usually happens when such appearances of spatial relations occur.

In the case of GOIs, the appearance usually comes from the figures being at different distances to the subject. For example, in the Ebbinghaus illusion, all but the central circles divide into rings of otherwise identical circles. However, in one ring, all the circles are bigger than the circles in the other ring. Usually, this apparent difference is due to the first ring being *nearer* than the other ring. When this usually happens, each ring appears as it does because it is at a different depth to the other ring.

If that is right, then if something appears to be with one of the rings, it appears to be at a different distance to the other ring. In this illusion, each of the central circles appears to be with one of the rings. As such, each of the central circles (along with its rings) appears to be at a different distance to the other central circle (and ring).

This gives us the difference in size. Each circle appears to be a different distance. However, it appears to occupy the same visual width and height; if you move one circle only through the width and height dimension so that it covers the other, they exactly overlap. If we ignored differences in the third dimension, we might say they occupy the same region of two-dimensional space. However, once we bring in the appearance of a third dimension, one circle appears further back than the other circle. Thus one circle appears to be bigger and so to occupy a larger but farther off region of two-dimensional space.

Where could anosognosia come into this? It cannot come into the apparent difference in distance and depth. If there is such an apparent distance and depth, then it cannot be a merely incomplete perception of their actual depth. There is no difference in distance and depth. Both circles are equidistant.

As such, GOIs are not like mirages. They cannot be a difference between apparent and actual depth that is due to some *hidden* depth (complex or otherwise). There is no difference in depth here. There appears to a difference. This appearance is greater than the reality. As such, we should think of GOIs as illusions. We cannot re-interpret them as anosognosia.

Perhaps one method of avoiding an interpretation of illusion is to argue that, in the case of GOIs, that there is no phenomenology or appearance of depth (similar to the veridicalist approach to temporal passage (see the next chapter)). If that is right, then we could argue that the errors in such phenomena are only due to a failure in judgement, while the actual depth is hidden from us. That is, we are anosognosic of any actual depth and this leads to such errors. In that case, we use our perception of the non-depth properties—such as the relative sizes of most of the things we see—to judge the depth of what we perceive.

Unfortunately, I do not think this method works.

First, appeal to non-depth properties is not enough. The non-depth properties may be sufficient to give us judgements of visual depth. However, as in the example of Kölers' moving lights, non-depth properties can give an appearance of depth (along with other non-depth properties).

We need to say that, in this case, the non-depth properties do not give an appearance of depth.

This might be true for GOIs. I must confess that, despite the plausibility of Gregory's account, I am not sure if the figures in GOIs seem to be at different depths. This may be due to my believing other things (such as they are actually locations). However, given this account, if they do not appear to be at different distances, then they should not appear to be different sizes.

For GOIs, it may be false that an appearance of depth plays a role in the apparent size (in support of this, see Newman & Newman 1974). The figures do not appear to be at different distances. Does this mean the erroneous experience is anosognosia? For example, is it anosognosia of the actual depth of what we perceive, and through that, of the actual size?

I think not. Even if the figures do not appear to be at different depths (or even the same depth), they do appear to be different *sizes*. This is the original phenomenon needing explanation.

A final suggestion seems most plausible to me based on my own experience (others may differ in this). There is no appearance of depth in cases of GOI. However, in the perceptual process, at some point, there is some kind of unconscious judgement of depth. This judgement of depth influences the process that results in an appearance of size.

Others have developed this idea much more thoroughly than I wish to; it is part of Gregory's account (and of others cited earlier; also McCauley & Henrich 2006). I only bring it up to note this: This is not an account in terms of anosognosia. That kind of error is insufficient to account for the erroneous appearance of size. Perhaps we are unaware of the judgement of depth, and we might consider this anosognosia (in this case, of subjective activity). However, there is still the appearance of size.

We cannot explain this by anosognosia of something subjective. The size is of something objective; anosognosia of that size is only a failure to perceive all or some of something objective (and a lack of awareness of this failure). This might be a failure to perceive the whole object or the distance between the subject and the object. However, none of these is a *judgement* of depth.

By explaining apparent size as resulting from unconscious judgements of depth, we explain the experience in terms of a judgement and so something subjective. This is an account of the appearance of size only in terms of something subjective. This is an illusion, not merely ignorance of the limits of our perception. As such, it cannot be anosognosia.

8.3. Conclusion

This chapter discussed mirages and the erroneous experience of depth, attempting to understand them in terms of illusions and anosognosia. I argued that many are due to hidden depth in perceptual experience—i.e., cases of anosognosia. However, we cannot explain all the phenomenology of depth like this. In particular, where we posit apparent depth for GOIs, we cannot appeal to a failure to perceive hidden depth.

In this discussion, I have ignored one important feature of any such theory of perceptible depth (or indeed of any spatial properties). I believe that there is good reason to do so in this book. However, I need to mention it in passing.

In claiming that our experience of depth goes wrong, there is an assumption that there is real depth against which we can compare appearances. This raises a question: What defines that real depth? Is there some special allocentric or absolute coordinate system with which one can define real visual depth and judge apparent depth against it? If not, no matter how intuitive it may be, it is not clear if we can say that the apparent depth is mistaken. There is nothing available for evaluating the appearances.

I have already suggested that a determination of real visual depth requires the eye but does not require other parts of the subject, such as the body or brain. However, I am not sure the eye is sufficient for determining real visual depth.

Unfortunately, an account of what is sufficient must be left wanting. Such an account moves the discussion towards pure questions in the philosophy of space and so too far from the focus of this book.

To close, I develop an account of perceptual experience in terms of relations to things at other times and distortions of depth later when I speculate about the possibility of eliminating hallucination. Before doing that, I must discuss another important subject. This is the relationship between the philosophy of time and the constitution and content of the perceptual experience of time.

Note

1. There is some research about the relationship between (binocular) rivalry and experience in time. See Engels et al. (1999).

9 Temporal Experience

The last few chapters have applied thinking about the temporal structure of perceptual experience to the perceptual experience of space. I argued in those chapters that, if we want to be naïve realists about space—i.e., we want to preserve the phenomenology of spatial perception in our ontology—then we should be tenseless theorists (requiring only commitment to anosognosia) and not tensed theorists (requiring commitment to illusions), and we should be eternalists and not presentists (requiring commitment to hallucination).

This current chapter moves on to temporal experience, in particular, perceptual temporal experience. In this chapter, I consider perceptual temporal experience under two headings:

(1) *The perceptual experience of time*: The perceptual appearance or phenomenology of time or time as *content* (again, depending on how one uses such terminology).
(2) *The structure of the perceptual experience of time*: How experience is structured or constituted in order for there to be such temporal experience.

Note here that the chapter continues to focus on perceptual experience. For reasons discussed in this chapter, one may not wish temporal experience to be merely a form of perceptual experience.

Before getting into the discussion, I must put one possible difficulty aside. It may seem that a chapter on temporal experience should have come earlier, that temporal experience is the most important kind of experience to consider in thinking about experience and the philosophy of time.

However, the subject of this book is not only about temporal experience. It is not only the experience *of* time. The subject is experience *in* time. This chapter follows discussions about space because, to understand how we can apply considerations about being in time to the experience of time, it is helpful to consider them in a situation where there is less risk of confusion about the possible relationships between appearance and reality. Now that we have discussed time as a constituent of, as structure in, spatial experience, it should be easier to conceive of this role as distinct from temporal content. As such, in this chapter, I apply considerations about experience in time to temporal experience.

It may be an open question that temporal constitution and temporal content are distinct. Naïve realists, the preferred position about perception here, do not want it wholly distinct. However, that does not mean we can assume that there is no distinction. As with space, how it might be distinct, and what this means for different kinds of erroneous temporal experience, is something we must consider from different positions of time. As with space, it is possible that certain temporal experiences must be prone to error, whereas others may not. The argument of this chapter is that, even with temporal experience, we should not assume which experiences are erroneous and which are not without thinking about the influence of the philosophy time in our assumptions.

It is also important to note that, although this book does not focus on temporal experience, the subject of temporal experience can matter to the subject of more general experience. I accept that there is something temporal in the structure of *what* we experience. Along with being its own content (e.g., the experience of change, persistence, remembering, anticipating), time appears to be fundamental to other kinds of content as well, being part of or contributing to the phenomenology of other things.

This is because everything held to be experienced in the world seems to be something that occurs in time (and, in some cases, over time). For example, when there is an appearance of a coloured thing, a changing process, a tone, or an entertained thought, it appears as if the thing has its colour *at a time*, the process is changing *over a time*, the tone sounds *over or at a time*, and the thought is entertained *over a time*.

Remove time and perhaps we can still have instances of colours, things, processes, changes, tones, and thought. I do not know. As with mystical or divine experience, I have never experienced such non-temporal things. I also am not sure what it would be like to have such experiences. Such experiences would be of a colour outside of time, a thing existing or a process changing beyond time, hearing an atemporal tone, or thinking a thought which is neither brief or long, before anything else or after anything else (all of these are temporal features of the thought).

This fundamentality of time in experiential content gives a reason to be concerned with erroneous experiences of time.

In the chapter on illusion, it was stated that an experience of F can be erroneous (illusory, hallucinatory, anosognosia) if (a) F depends on G and (b) there is no G. For many kinds of experience, some kind of temporal property or relation is the 'G' here, the property on which other properties depend. We may have experiences which do not require time, but not many and not obviously. For example, perhaps we can experience mathematical entities as being timeless (I don't know), but the experience of doing mathematics is not of something timeless—e.g., long division by hand requires an experience of doing one thing after another.

If our experience of time is untrustworthy, it makes our experience untrustworthy in a general and fundamental way. If something we experience as having a particular temporal structure does not have that temporal structure, then we no longer can tell from experience much of what determines the structure and properties of that thing. For example, if I seem to see a ball move through space, but it is not moving at all, nor does it even have temporal properties such as persistence, it is hard to consider it to be a particular physical object, never mind anything so specific as a ball.

As such, it is important to ask how temporal experiences corresponds to the world. There are three questions considered in this chapter:

(1) What kinds of temporal experiences are there, if any? Do we experience change, duration, simultaneity, succession, temporal order, and passage? Do we only experience some of them, and not others?
(2) What kinds of erroneous temporal experiences are there, if any? Can we have temporal illusions, temporal hallucinations, or temporal anosognosia? Given time's relationship to other properties and objects, the possibility of these errors for time can make it more likely that these errors are possible for experiences of many other properties and objects.
(3) How are temporal experiences structured and constituted? If we must have certain constraints on the structure and constitution of temporal experience, there may be similar constraints on the structure and constitution of experience in general.

These questions depend in part on how we think about time. However, in this book, I do not consider them for general temporal experience. I consider them only for the perceptual experience of time. This book is about perceptual experience because of its epistemological role and its particular nonrepresentational, non-intentional, realist character. Other kinds of experience are far less obviously possessive of these features.

Other kinds of experiences which particularly concern time—e.g., memory experience and anticipatory experience—even if possessive of success conditions (e.g., your memory experiences can be accurate and your anticipations can be met), seem to be only representational or intentional experiences, such as imagination. They do not seem to be constituted by what is experienced.

As such, even if we may want the world to match memory experience or anticipatory experience, we would not hold that the experience itself contains part of that world. That the experience occurs would not be enough to justify holding that its contents occur.

Perceptual experience is different, as indicated by naïve realism. It seems to have the thing experienced as a constituent. As such, if there is a perceptual experience of time or some temporal property or relation, then it seems to be in some way a constituent of the perceptual experience itself.

My aim in this chapter is to explore how we might have time or some temporal property or relation be a constituent of the experience. To do this, we must consider erroneous temporal experiences. As with many experiences discussed here, the best way for a naïve realist to account for such erroneous experiences is to appeal to an anosognosia of time lags between such experiences' constituents.

9.1. Perceptual Experiences of Time

In the literature, it is common to talk about time perception, time consciousness, and temporal experience. One might assume that these terms are synonyms of each other, and we can substitute each one for another. That is, every instance of (a) time perception is an instance of (b) temporal experience and every instance of (b) is an instance of (a).

However, there are reasons to hold that (a) can come apart from (b) (and vice versa). As already discussed, not all experience is perceptual experience, and many of the ones that are not are particularly temporal. Memory experiences and anticipatory experiences are not perceptual experiences.

As such, when one talks about perceptual experiences of time and general temporal experience, one must be careful to acknowledge that they might not be the same thing. Not doing so can lead to assumptions and confusions about what is being explained and how to explain it. For example, one may provide answers that are inadequate and raise objections which are irrelevant.

Here are some examples of cases where this can happen, where one might treat the two as identical or, if treated differently, the difference is not significant.

9.1.1. 'Time Perception' Is Only Quantitatively Different to Other Temporal Experiences

We may assume that the phrases 'temporal experience' and 'time perception' both refer to what I refer to here as 'temporal experience'. In that case, there are two possibilities:

(a) *We hold that both phrases refer to experiences that are explained how all cases of possible temporal experience are explained.* This includes experiences of time that we would never seriously refer to as perceptions of time—e.g., the experience of the duration between now and the death of the sun. This is not even an anticipatory experience. If it is any kind of experience, I assume this is an imaginative experience of duration.

(b) *We allow a difference between time perception and temporal experience that is only a matter of quantity or magnitude.* Perception is only of present things; we do not perceive the past or future. Thus perceptual experiences of time are experiences that are of time *in the present.* That is the only difference between those experiences and other kinds of temporal experience. The perceptual experience of time is of the quantity of time that is present, contrasted to other durations only by its brevity. Yet quantity is not important to explaining an experience. As such, as with (a), we explain it in exactly the same way as we do experiences of other kinds of temporal experience.

If we follow (a) or (b), then we get the following situation: We must account for perceptual experiences of time in some common way to how we account for memory experiences or anticipatory experiences. We might even need to account for them the way we account for experiences of durations beyond our own lifetime (such as the far future).

If someone thinks this way, then, because they include obviously intentional experiences—e.g., of future or imagined times, they appear justified in holding that time and temporal properties (relations, etc.) have to be something we can only intend, imagine, or, in some way, think about. It is intuitive to assume that the vehicle of an imagined Big Bang does not extend back through time to the Big Bang. The duration between the subject and those events (along with the events) is on an ontological and constitutional par with conceivable unicorns and the surfaces of inhabited alien planets. Even if time is real, we must only have some kind of intentional attitudes towards it.

If that is right, references to 'time perception' are misleading. We do not perceive the duration or temporal properties of present things. What applies to 'time perception' is what also applies to experiences of the duration stretching to the Big Bang and all other experiences of time. The only difference is quantitative—i.e., it is only a matter of how long the particular duration that we experience is.

Thus given this identification of perceptual experiences of time with other kinds of experience, the perceptual experience of time has to be intentional.

I do not think the perceptual experience of time has to be of something merely intentional. I do not even think that we should even *want* it to be merely intentional. Like much of what we perceptually experience, we should want it to be real. In addition, I think that we can have it be real.

9.1.2. Perceived Change and the 'Specious' Present

There is a phenomenological reason to resist identifying the perceptual experience of time with other kinds of temporal experience. This difference is not just a matter of the quantity of duration. It involves the experience of present duration or change. This is the 'specious present' discussed in Chapter 6.

One may hold that, with respect to this duration (the time of the specious present), a perceptual experience time is as consistent with naïve realism as a perceptual experience of space. For example, the constituents of our perceptual experience spread through space and appear spread through space; similarly, the constituents of our perceptual experience spread through time and *appear* spread through time. Just as there is a perceptual experience of the different spatial properties of (or of spatial relations between) what we experience, so there is a perceptual experience of the different temporal properties of (or temporal relations between) what we experience. In having a perceptual experience, we are perceptually related to things at different times.

If there is a perceptual experience of change, then we have reason to think that the difference between time perception and general temporal experience is not merely a matter of quantity. It is a matter of quality. There is a qualitatively different phenomenology, or even a categorical difference, between two kinds of temporal experience.

Here is how I divide the two by way of their extreme forms: that of the duration stretching back to the Big Bang and that of the duration of perceived change.

- One (of the Big Bang–related duration) is like imagination or memory or anticipation. I can conceive of them as merely intentional and thought-like, such as imagination.
- The other (of perceived change) is perceptual. The latter is not merely intentional or thought-like, unlike imagination.[1]

If this is right, time perception does not appear to be a mere case of temporal experience like other kinds of temporal experience (such as that of past or future life events, or the Big Bang). It seems to be a perceiving of temporal properties.

However, one might think the opposite holds about the relationship between time perception and temporal experience. One might think that, whenever someone refers to temporal experience, they mean time perception. This can raise a different set of confusions to those from thinking time perception is just temporal experience.

Once one holds one or two commonly held positions, here is one way to get confused. One holds that (a) perception always refers to present things and (b) present things are punctal.

Holding these, one might be puzzled by how there could be temporal experience other than the experience of whatever temporal properties are compatible with a punctal present. Consider the duration to the Big Bang. You do not experience that. Consider the duration between you and any event in the past or the future. You cannot experience that duration because no such duration is present.

Therefore, the only temporal experience would be experience of time in the present. However, as already discussed, we do have such experiences—of non-present events—if we have experiences of remembered or anticipated events. Remembered events are past; so if we experience remembered events, then we experience past events.

For these reasons, I take the following position: time perception and temporal experience refer to related but different things:

(i) The perceptual experience of time or time perception refers to things like perceived change, time in the experienced present, the specious present, etc.
(ii) Temporal experience refers to a general experience. It may cover time perception (where the perception includes perceptual experience), but 'time perception' is qualitatively different to it. It also concerns all other kinds of experiences involving time. It includes whatever we obviously remember and what we anticipate. If we have experience in some cases of imagining, it includes whatever it is we call the 'experience' of durations over eons, etc.—i.e., imagined time.

What I am most concerned with here are experiences and models related to (i) with time perception. In this book, I am not so interested in more general phenomenon of temporal experience. Again, the reason for this focus is that, unlike imagination and cognitive phenomenology, the phenomenology of perception invites resistance to merely intentional or thought-like models of perception. It does not seem to involve only something intended or thought. It seems to involve something independent of us, real, in the world. For this reason, it supports one ontology over another—those that correspond to the phenomenology, something absent in—e.g., imaginative experiences.

As such, the perceptual experience of time is ontologically interesting and so interesting to the ontological debates about time (memory experience is also interesting to some extent in this way).

This is why I argue also that (a) the case for erroneous perceptual experiences of time and (b) the plausibility of different models of perceptual experience of time's constitution depend in part on one's views on time.

At this point, it may be useful to discuss the phenomenology of the perceptual experience of time on its own. Once we have detailed that, we can move on to the veridicality of such perceptual experience or its relationship to the philosophy of time. However, to do so would extend this work beyond its intent. There is a lot of literature out there about the phenomenology of time perception.

Instead, I discuss all three together. In the next section, I discuss what kinds of veridical and erroneous perceptual experiences of time there can be, given different concepts of time.

9.2. Kinds of Erroneous Temporal Experiences

If we have experiences of time, what kinds of erroneous temporal experiences can there be? It may seem we need only look at the empirical evidence and then decide. As with perceptions of space, we can do that once we settle the more general questions of what we seem to experience and what is real about time.

Here I argue that there are no temporal hallucinations, there can be temporal illusions, and there can be temporal anosognosia. However, we can better understand some cases of illusion as anosognosia.

9.2.1. *Temporal Hallucinations*

Except in the sense of a hallucination being a complex illusion (as I discuss in Chapter 4), there are no temporal hallucinations. At least, if one asserts that there are temporal hallucination, they must do so based on implausible premises. For temporal hallucination, there would appear to be time that is an object. Yet time does not appear to be an object.

There may be philosophical reasons for holding that time is an object. For example, philosophers who are *substantivalists* argue that time is a *substance*. For example, some theorists argue that questions about curving of space-time in general relativity (Le Poidevin 2003) and the simplicity of constitution of things in space-time (e.g., Sider 2003) may better explain that space-time—and by extension, space and time—is a substance. If those arguments work, and one holds that substances are objects (although, this is not obviously true), then time is an object.

However, it is generally accepted that a motivation for its being a substance (and so an object) is not its *appearance* as a substance (or object). The arguments are for a time that is independent of changes and persistence in things. Further, it is independent of our experience, as it is generally accepted that we are only aware of time in the persistence and change of things (e.g., for discussion, see Shoemaker 1993). Thus if there is such an independent time, it is not apparent to us.

If that is right, then a substantial time, a time that is an *object*, is not apparent. Whatever errors there may be concerning time as an object, they are not hallucinations. Rather, if one wishes to argue that we can hallucinate time, one must argue that we do experience substantial time, or that we mean by 'time as an object' something other than it's a substance (I leave others to decide what that means).

9.2.2. Temporal Illusions and Temporal Anosognosia

There are many cases of erroneous experiences described as temporal illusions in the literature. They include the *phi phenomenon*, unusual experiences of duration (e.g., in strenuous circumstances or the 'oddball' effect—e.g., Power (forthcoming)), and perhaps the relative timing of things, such as the flash-lag illusion. The concept of a temporal illusion is somewhat plausible, unlike hallucinations. Even so, some of these are not necessarily temporal illusions.

A temporal illusion is an erroneous experience of temporal properties and relations. As with hallucinations, for illusions, we need to hold that temporal properties are apparent to us and that their appearance is in the absence of their occurrence.

This puts a particular restraint on the class of experiences that we may call temporal illusions. Recall in the discussion about the conditions for illusion, it was noted that it does not follow from an illusion of some G which depends on F that there is an illusion of F, even if F is apparent. Here this means a temporal illusion does not follow from there being an illusion of something that *depends* on time. Nor does the fact there are some kinds of temporal illusions mean that there are other kinds of temporal illusions. It does not follow from appearance of F and G, and not (G), that there is an illusion of F.

First, temporal illusion is not merely an illusion of something that depends on time. This may be obvious, but it is worth making it explicit. There need not be a temporal illusion because there is an illusion of motion, change, or persistence. Such properties may need time (e.g., there is no timeless motion or change). However, there can be erroneous experiences of these without erroneous experiences of the temporal properties or relations on which they depend. This is because the appearance of *other* properties on which they depend could be mistaken.

For example, motion is change in something's spatial location. The temporal order and duration of such a change could be right, yet the spatial properties are mistaken. Watching a film, it appears to us as if a car moves across a desert. We do perceive a change—we do see visible events in the order in which they occur (one frame after another). What we do not see is a car changing location. There is no car in the film theatre.

Another example is *apparent motion*: Motion requires a single thing that persists from one location to the other. Two separate events at separate locations A and B briefly appear in succession, and seem to be *one* thing moving from A to B. There is not such a single thing here and so there is no motion (similar errors underlie the cartwheel effect). However, there is no reason from just this to deny the temporal properties, such as the sequence's duration, temporal order, or even that there is a change (although, as discussed later and already by Dennett 1991; Bardon 2013, and to a weaker extent Paul 2010), apparent motion *might* motivate that).

Second, an error of *some* temporal properties is not an error of *all* temporal properties. For example, perhaps there are illusions of simultaneity; perhaps we have good evidence that some phenomena are due to or themselves instances of illusory simultaneity. This does not mean that one now has evidence of illusions of order or duration. They are different and often exclusive properties. As perceptions of one are not perceptions of the other, so failure and illusion of one is not failure and illusion of the other. To have it that an error of one temporal property is an error of another temporal property, we need it to be that there is some dependency between them.

For example, if an event E's pastness depends on E's relation of being earlier than some other event E', then if it is not earlier than E', it is not past. If it should appear to be past, then this is an illusion of its pastness.

The same follows with temporal anosognosia. There may be a non-appearance of some actual instance of a property or relation (or object), something which depends on time. Yet we are not anosognosic of temporal properties and relations (or object) unless the appearance of the temporal needs the appearance of the hidden thing.

What could be an example of this? As discussed, we only experience time as properties and relations, and so we only experience time as it qualifies or relates concrete particulars and events. In addition, we are not aware of any actual particulars or events, but they are there. In that case, we are anosognosic of time (as well as the particulars and event). (Although, if we do not experience particulars and events, it is hard to imagine what *is* apparent.)

Here is a better reason to be anosognosic of time. We are anosognosic because the apparent temporal properties and relations of particulars and events we experience are the negation of other temporal properties and relations. That is, if these properties and relations are not instantiated, then the other temporal properties or relations are instantiated.

We can then get anosognosia as follows: The apparent temporal properties do not belong to the things we otherwise experience. Their opposite number do belong to what we otherwise experience. In that case, we are anosognosic of the temporal properties of what we experience.

There are many instances of this, especially with respect to simultaneity. Take the experience of distant things at different times. What we experience given tenseless theory is at a different time to us (and our subjective activity). However, it does not seem that way. It seems to be simultaneous with the subjective activity. That is, in having a false appearance—what we might call a temporal illusion—of simultaneity, we have temporal anosognosia of duration.

This opens a last possibility: In some cases, we can understand temporal illusions of one type of temporal property or relation as temporal anosognosia of the negation of that property or relation. We can hold, for example, that all there is to an 'illusion of simultaneity' is anosognosia of temporal difference or duration.

Let us consider these distinctions as applied to specific examples of erroneous temporal experience.

9.3. Philosophy and Errors of Time

If something moves past too quickly, then I see it as a blurred object. This blurred object occupies different locations at different times but seems to occupy those locations at the same time. We can describe this common perceptual experience of something in either of the following ways:

(a) It is composed of stages of a rapid motion that appear simultaneous, yet are not simultaneous.
(b) It is composed of stages that do not appear to be at different times, yet are at different times.

An uncommon pathological case of this is *akinetopsia* or *motion blindness*. One is unable to perceive the continuous motion of an object most people can see. One is only able to perceive different stages of it frozen in place—e.g., filling a cup with tea, the level of the cup appears unchanged during a period, and then, suddenly, the cup is overflowing (e.g., Kelly, *op.cit;* Le Poidevin 2007).

Although it is usually only applied to the uncommon experience, we can refer to both as a kind of *motion blindness*. As might be obvious, it is a variation of change blindness, discussed in the chapter on anosognosia.

One way to look at such motion blindness is as a temporal illusion. For some object, there is an illusion of its stages being simultaneous. Yet, as with change blindness, another way to interpret this motion blindness is as its name implies. The subject cannot see the motion; they cannot see it moving from one place to another place. They are *blind* to the difference in the times in which the object occupies each location. Either they don't experience the duration between the stages or, with akinetopsia, during the duration of the change, they are only aware of a limited number of the stages—e.g., the water first pouring and then overflowing. This is similar to a subject being blind to the difference in the location of two occluding objects, or two objects too small and too close together for them to be perceptually distinguishable.

So what sort of error is this? It varies with our view on time.

9.3.1. Blindness to Multiple Times

Eternalists can say that they do not experience the difference in the times of the stages. This is a failure to experience a difference between real things. We are anosognosic of the duration. Furthermore, if relativity is true, then the apparent simultaneity fails in another way: It fails to appear simultaneous relative to frames of reference. This is anosognosia of the relativity of what we experience (we return to that kind of error later).

Presentists should not say that we do not experience a difference in the times of the events. This is not a failure to experience a difference between real things. One of those things—whichever event is not present—is not real. Furthermore, these are perceptual experiences—experiences of present things. With presentism, whatever it is we experience, if it is real, then it is present. Thus, there is no duration between any of its stages or parts. As such, there is no error here of their temporal relationships—i.e., the apparent simultaneity of what is experienced corresponds to *real* simultaneity.

However, since the stages of what we experience as simultaneous are not simultaneous, then we cannot be seeing those real stages. Thus the appearances are mistaken in at least one of two ways:

(a) *What we see is not the set of stages that we seem to see*. It is a set of other present and simultaneous events (being simultaneous, I am not sure it is right to call these 'stages').
(b) *Only one of the stages we seem to see is real*. The present stage is real. All other stages are unreal.

Either case is no experience of temporal anosognosia. Under presentism, we cannot explain an experience of something (such as stages at different times) as merely failing to see what is real (such as one of those stages that are at non-present times). What marks the difference here makes one of the stages unreal.

With respect to erroneous temporal experience, under presentism, this experience is at least one of the following:

(a) *A temporal illusion*: Of at least one stage's presentness—and, through it, of its existence. As such, we are also hallucinating the stage.
(b) *Not a temporal error at all*: It is an error of *what* we experience. We do not experience the non-present stages, because they are past. We can experience other things, because they are present (e.g., sense data).

Tense theory: The problems for tense theory are not as great as presentism, but are also not as simple as tenseless theory. The stages can be real. However, their mutual presentness cannot. All but one of the stages must be past or future. Seeming present, they are illusorily present.

Some might think that there is another approach for tense theory. One might hold that an 'illusion of presentness' is merely anosognosia of the other tenses—e.g., of the past, the future (if we experience future events), or both. This is to hold, for example, that when two stages appear to be present, one is past. That stage's appearance of presentness is merely a blindness to its pastness.

However, this approach depends on metaphysical assumptions, ones that tense theorists do not have. For tense experience under tense theory, illusory presentness cannot be mere anosognosia of another tense. Tense is a fundamental property of things, not merely the absence of some other property. What is present, past, or future is fundamentally and absolutely so.

However, for tense experience under tenseless theory, it can be merely the absence of something else (such as duration).

This discussion concerns motion. However, what about other perceptual experiences involving time? What can we say about the perceptual experience of change, passage, and persistence, under different theories of time?

9.3.2. Perceptual Experience of Tenseless Change

There are many erroneous experiences involving change—e.g., change blindness (including motion blindness), motion after-effect, the feeling of moving when you are not moving, the wagon-wheel effect. We have already discussed change and motion blindness. If we have such erroneous experiences, they are anosognosic experiences of change.

Does a difference in one's philosophy of time make any difference in the interpretation of other erroneous experiences of change? It does in two ways: First, by making a difference in the range of what counts as change, and so what counts as an illusion of change. Second, by making a difference in the interpretation of experiences of change as illusory (or generally erroneous).

As discussed in the chapter on philosophy of time, there are two general concepts of change:

- *Temporal passage*: The change in tenses or A-series positions—e.g., the death of the sun changes from being future, to being present, to being past.
- *Tenseless change*: The change in qualities over B-series positions—e.g., an apple is raw before the apple is ripe before it is rotten.

If we do not experience change where there is change, then this is anosognosia of the change. If we do seem to experience change where there is no change, this is illusion of change. Yet given our theories of time, what sort of commitments do we have to change?

Do we perceptually experience tenseless change? We do if we perceptually experience motion. Motion is some A being at one place *and then* A being at another place. If 'and then' is cashed out in terms of tenseless relations such as 'before', this is sufficient for motion itself. More generally, if some A is F *and then* A is ~F, this is tenseless change.

If we perceptually experience such change, then this is consistent with tenseless theory. There is no obvious reason to hold that such an experience is an illusion of change.

There is some cause for holding that we perceptually experience such tenseless change. We perceptually experience *some kind of change*. Again, with motion: wave your hand in front of your face. Alternatively, listen to the rise and fall of an ambulance siren, the wind rattling the window, the ebb and flow of your back pain—anything you can perceptually experience as changing. Then, from that, we can construct an argument for that change being tenseless (the following is adapted from Power 2013).

Argument for Perceptually Experienced Tenseless Change

(1) The change we perceptually experience is either tensed change or tenseless change.
(2) Any change is being (F *and then* ~F).
(3) Everything we perceptually experience is present.
(4) When something is present, it is not (even partly) past or future.

A. Putting some of these together:

(5) (From 4) When a change is present, it is not (even partly) past or future.
(6) (From 3) Any change we perceptually experience is present change.
(7) (From 2 and 6) Any change we perceptually experience is present *being* (F *and then* ~F).

B. What presence means for perceived change:

(8) (From 4 and 7) When a *being (F and then ~F)* is present, it is not (even partly) past or future.
(9) (From 6 and 8) We perceptually experience *being (F and then ~F)* that is not (even partly) past or future.

C. What tensed change needs:

(10) If changed is tensed change, then 'and then' is either between a past event and a present event, a present event and a future event, or a past event and a future event.
(11) (From 2 and 10) If change is tensed change, then it is being (past event and then present event) or being (present event and then future event) or being (past event and then future event).
(12) (From 11) If change is tensed change, then *being (F and then ~F)* is at least partly past or future (either F or ~F is past or future).

D. (9) denies that we perceptually experience change that is past or future.

So

(13) (From 9 and 12) We do not perceptually experience tensed change.

E. (1) holds that either we perceptually experience tensed change or we perceptually experience tenseless change.

Conclusion. (From 1 and 13) We perceptually experience tenseless change.

Other responses are possible here:

We could deny (1), that we perceptually experience change (as Plumer 1985 does). However, that is not the general contemporary view.
We could deny (2) and hold that we some things we perceptually experience are past or future. I return to that later.
We could deny (3), that change in something is it being F at one time and *then* ~F at another time. Instead, change is something else that can be *all at one time*. For example, we might say that change is a state (at one time and so present) of being F and *becoming* ~F. I return to that later.
We could deny (4) and hold that the present can be partly past or future (one might think of this as an instance of the 'specious' present). Thus, an event can be past when it is present.

This may be an approach, but it is not one tense theorists can agree to. Tense theorists distinguish locations in the A-series by tense. If the exact same location in time could be said to be the present and the past (or future), what prevents it from being two separate locations is indescribable by the A-series. This is especially exasperated with the present, of which there is only one in the A-series. Here, any location can also be present, as its location in the past or future does not prevent it being so.[2]

Returning to the argument, does it mean that tense theory requires such a perceptual experience, being of tenseless change, to be an erroneous experience of change? Yes, but it is not so bad. It is just a case of anosognosia.

Recall that change in tense theory is temporal passage: An event being in one location in the A-series then being in another location in the A-series—e.g., an event being future—then being present, then being past. Mere tenseless change is not this passage. However, we derive tenseless change from passage.

Tenseless change is something being F and then being ~F. Modified (or corrected) for tense theory, this tenseless change is something (a) being presently F and being futurely ~F, then (b) being pastly F and being presently ~F, then (c) being more pastly F and being pastly ~F . . . etc.

What happens in perceiving tenseless change is we fail to detect the tenses of the different stages of the change. If we perceptually experience tenseless theory, then it just seems to us as if there is F and then ~F. We do not detect F leaving the present and becoming past, or ~F entering the present from the future, etc. We just detect the F being present and then ~F being present, without the appearance of either property instantiation changing tense. That is, we experience each stage as present in order but fail to experience each stage's *real* change—the change in its *tense*.

However, there is a more problematic error for tenseless theorists and the appearance of tensed change. If we perceptually experience temporal passage, denying it—as tenseless theorists do—is not obviously a case of anosognosia. It seems to require an illusion. I suggest a way to avoid that conclusion, although not to wholly eliminate error. It depends on how we understand what it is we fail to experience.

9.3.3. Temporal Passage

As discussed, it is possible to deny that we perceptually experience tenseless change. We can deny the premises that we perceptually experience only the present. Instead, we perceptually experience a period of time that includes the present, but also includes some of the past and future. This is why the 'specious' present is so-called by James: It is not one moment in time, and so is not really the present (e.g., James 1890; for discussion, Power 2012).

As previously stated, I think we should resist that for perceptual experience. It can be of the present, given how tenseless theorists conceive of the present. However, one might provide a convincing argument even to tenseless theorists that we must experience change as partly past or future (although I have no notion of how that argument might go, perhaps models might play a role—see the following).

If that is the case, then we perceptually experience temporal passage. We perceptually experience events being future, then present, and then past. We perceptually experience things changing by their *having been* ~F and *being* F (or their *being about to become* ~F and their *being F*). Some things we perceive *appear* past (or *appear* future). If naïve realism is true, then past (or future) events partially constitute perceptual experience, and we are perceptually related to past (or future) things.[3]

If one is a tenseless theorist, one denies this temporal passage. Indeed, recently tenseless theorists have argued that, even if there is temporal passage, and independent of a commitment to tenseless theory, we do not or cannot experience temporal passage (e.g., Prosser 2000; Prosser 2012; Miller forthcoming).

As such, if we experience temporal passage, then this looks like a simple case of temporal illusion. There is the appearance of passage, and there is no passage. Furthermore, it does not look as if it can be anosognosia. This is an appearance of tense change. Tenseless change is not the same as tense change. Furthermore, the two kinds of change happen to different things: Tenseless changes happen to things; tensed change happens to events (or to the present's relation to events).

However, it is not clear that tenseless theorists must hold that there is an illusion of tensed change in erroneous experiences of tensed change. Those that do grant that there is such a phenomenology yet deny there is any passage are *illusionists* about temporal passage. However, one can also deny there is even an appearance of temporal passage at all (a position that some theorists call *veridicalism*) (Deng 2017; Miller 2017) and Torrengo calls *deflationism* (Torrengo 2017).

This approach may be one way to dissolve phenomenological worries about passage. However, here I am interested in the situation where we are convinced that there is an appearance of passage. In that context, I suggest a third answer: The subject is anosognosic of what is experienced in such a way that there appears to be passage. To do this, we must extend the discussion of experience beyond discussion about perceptual experience. Thus this goes beyond the theme of this book. However, it is worth briefly diverting to it; perceptual experience—indeed, of change—still plays a crucial role in this account.

The account turns on two ways of classifying properties discussed in the chapter on relativity. This is the classification of properties as relative or absolute. In addition, it turns on an important feature of change that is common to tenseless change and passage.

9.3.4. *Temporal Order*

Common to change in both its tenseless and tensed character is the succession of the stages constituting the change. If there is a change because there is F *and then* there is not (F), the 'and then' indicates that there is a temporal *order* to there being F and there not being F.

In tenseless terms, this is

- F *before* not (F) . . . and that is different to . . .
- F *after* not (F)

In tensed terms, this is

- *Was* (F) and not (F). Or F and *will be* (not (F)) . . . and that is different to . . .
- *Was* (not (F)) and F. Or not (F) and *will be* (F)

That is, there is temporal order behind both kinds of change. The temporal relation between the stages of a change is asymmetric so that, even if everything else is otherwise the same (e.g., there is only F and not (F)), there can be a difference in the world just by each stage being a different relatum in such a temporal relation.

It might seem that, in having an experience of change, we also have an experience of temporal order.[4] Without the ordering of the stages, there is no change here. If that is right, how does that relate to the experience of change in general and to particular errors of this change?

Errors of this order undermine the experience of change in general. Other temporal properties depend on order. This makes it unlike other kinds of alleged temporal illusions, such as motion. Whereas an illusion of motion does not require that there is a temporal illusion of change, an illusion of order does. Once the order of a change is illusory, it is not clear that there is anything left to call change.

This is important because there seem to be reasons to hold that there are illusions of temporal order. As such, there may be reasons to hold that there are illusions of tenseless change, temporal passage, and anything else requiring order.

The cases of erroneous experience that theorists commonly advance as illusions of temporal order are the following: apparent motion (in particular, the phi phenomenon) and the cutaneous rabbit. I turn to them in more detail next. Before that, I want to mention three ways theorists have used them in the debate about the experience and philosophy of time.

Philosophers of time use them to undermine the possibility of temporal passage (e.g., Paul 2010; Bardon 2013). Others use it to support a conception of perceptual experience as being exhaustively representational (e.g., Dennett 1991; Grush 2005).

Paul argues that the 'flow' of apparent motion does not require real flow. It is something constructed by the mind, and a flowless state can represent it (and, if it is a state, it need not have a real *change* as a constituent). Similarly, the 'flow' of passage does not require real 'flow' and so does not require real passage.

Bardon argues for the more general point that such apparent motion can occur in the absence of any real temporal order. It can be *merely* apparent temporal order. As such, there is no need for temporal passage.

Dennett and Grush have a different motivation but it requires the same assertion of illusory temporal order. They argue for the 'multiple drafts' model, a model in which perceptual experience of a series of temporally ordered events is constantly revising moment by moment. At one moment, we have a perceptual experience as of events over a larger period—e.g., of A, then B, then C, then D. At the next moment, we have a perceptual as of events overlapping that previous period, and most of the same events, but not necessarily of the events in the same order—e.g., of B, then D, then C, then E.

There looks to be some justification for denying the appearance of temporal order. As said, the most notable examples are colour *phi* and the cutaneous rabbit. The common way of presenting them undermines apparent temporal order in two different ways.

Let us begin with the colour phi phenomenon. Here is one way the experiment might go: subjects are presented with one of the following sets of stimuli:

(1) A flash (meaning it stops before the next stimulus) of colour A at location S and, a moment later, a flash of the same colour at location S'.
(2) A flash of colour A at location S and, a moment later, a flash of a different colour B at location S'.
(3) A flash of colour A at location S and, a moment later, nothing.

Subjects then report what they experience.

Given (1), they report the motion of a single thing of colour A from S to S'.

Given (2), they report the motion of a single thing from S to S', a single thing which changes from A to B midway through.

Given (3), they report no motion of a single thing. They report just the flash of A at S.

Although (1) is an illusion of motion, it also seems to be explicable in terms of a failure to identify different flashes as belonging to different things. As such, we can interpret it as an illusion of the persistence of a single thing that moves. There is no need to talk about illusions of time and certainly not of temporal order.[5]

Why would anyone think it does lead to illusions of temporal order? The problem comes from the combination of (2) and (3). It comes from the appearance of A changing into B in (2), an appearance which is absent in (3).

If the second stimulus—B at S'—is presented, as in (2), then subjects report experiencing A changing into B. However, if the second stimulus is not presented, as in (3), then subjects do not report experiencing A changing into B.

This suggests that the occurrence of B (and, of course, A) is the cause of the experience of A changing into B. However, B appears to be later than the event of A changing into B. It appears to be the result of that change (and so to come after it).

Given this account, what we have is a case for saying that there is an illusion of temporal order.

Let 'an appearance of X' be denoted by 'app(X)':

(1) There is app (A, then A-changing-into-B, then B).
(2) Either

    (i) (A and B)
    (ii) app (A and B)
    (iii) (app (A) and app (B))
    . . . cause app (A-changing-into-B).
    So there is A, then B, and *then* app (A-changing-into-B).

(3) The time when app (A or B) happens seems to be identical to the time when A or B happens. So

    (a) There is A (or app (A)), then B, then app (A-changing-into-B), then app (B).

(b) There is A (or app (A)), then app (B), then app (A-changing-into-B), then app (B).
(c) There is A (or app (A)), then app (A then B), then app (A-changing-into-B), then app (B).

Given any of these options, there is trouble for temporal order.

If (3a) is the case, then it seems as if B happens after the change, and it does not. It happens before the change.

If (3b) is the case, then it might seem as if B happens in the absence of the change for a moment, and the temporal order is fine; that is the first app (B). However, then there's app (A-changing-into-B) and then app (B) again. Not only does B seem to happen after the change, but it seems to happen before the change.

Lastly, if 3(c) is the case, there is an appearance of A and then B, then of the change between them, then of B again. This is no help.

Taken as an exhaustive account of our possible relationships to what we experience, all these options spell trouble for there being an appearance of temporal order. As far as these options go, B happens before the apparent change—for there to *be* an apparent change, it *must*. However, B appears to come after the change (after all, it appears to be the *result* of the change).

The cutaneous rabbit is another example like this. Here subjects are presented with a series of taps along their arm. There are two scenarios:

(1) *The taps continue for only part of a period of time*—e.g., one on the wrist, two on the lower arm, and that is all. The taps feel like a series of disconnected taps.
(2) *The taps continue for the whole of a period time*—e.g., one on the wrist, two on the lower arm, one on the upper arm, two further up the upper arm. The taps feel to the subject as if they are and have been a 'hopping' of something along their arm, like a rabbit. That is, the sequence of stimuli feels like a sequence of *pairs* of stimuli rather than the disconnected set felt before it is broken off.

This case challenges temporal order because, again, when the tapping is broken off, the apparent order of the *same* events is different to the apparent order of those events when the tapping continues.

For example, midway through the experiment, the experience is of one tap (call it T1) *then* two taps (T2 and T3) *then* one tap (T4). However, afterwards, the experience is of two taps (T1 and T2) *then* two taps (T3 and T4). If these are the same taps in both cases, the *order* has to be in some way wrong for one of the experiences:

- For the midway experience, the apparent order is T1 *then* (T2 and T3) *then* T4.
- For the end experiences, the apparent order is (T1 and T2) *then* (T3 and T4).

Grant the differences between appearance and reality their extremes: For each mistaken experience, every temporal relation is mistaken (it need not be like that, but we gain nothing by being more particular). We then get the following:[6]

(1) *The midway experience is wrong and the end is right*: It is not T1 *then* T2, nor is it T3 *then* T4, nor is it T2 *and* T3. Leading to

   (a) There is no order between T1 and T2 or T3 and T4, yet there appears to be. These are *illusions* of temporal order.
   (b) There is order between T2 and T3, yet there does not appear to be. This is *anosognosia* of temporal order.

(2) *The end experience is wrong and the midway experience is right*: It is not T1 *and* T2, nor is it T3 *and* T4, nor is it (T1 or T2) *then* (T3 or T4). Leading to:

   (a) There is no order between (T1 or T2) and (T3 or T4), yet there appears to be. These are *illusions* of temporal order.
   (b) There is order between T1 and T2, and T3 and T4, yet there does not appear to be. This is *anosognosia* of temporal order.

Given these examples of actual experience, is there any way to preserve a naïve position about temporal order?

We can if we do the following:

First, we consider these cases in light of the account of naïve perceptual experiences of time.
Second, give up on something we have given up already, that what we perceptually experience is simultaneous with the subjective activity of our experience.
Third, we hold that we experience the temporal order of some of our *subjective activity* and not of events in the world.

The resulting account we give is of an erroneous experience, a kind of anosognosia. The position that we perceptually experience the underlying temporal order of the world, however phenomenologically justified, must be false. However, it is not a surprising erroneous experience.

It is a particular instance of a general kind of erroneous experience. Even without the phi phenomenon or cutaneous rabbit, we have reason to doubt that the apparent temporal of events is identical to the actual temporal order of events. This is because of what sort of events are apparent to us. These are external events, such as star shine, thunderstorms, and conversation.

Such external things are at a different time to their related subjective activity in us. However, if they are perceptually experienced together (e.g., I hear a conversation while looking at the stars), they appear to be at the same time as the subjective activity (e.g., my feelings about both conversation and stars). In addition, these external events are commonly at different times to each other (e.g., the star shine is at a different time to the speech around me). However, again, they appear to be at the same time as each other.

There need not appear to be any of the actual temporal order between these external events and our subjective activity (which they partially cause) or between the external events themselves.

Thus independent of what we might say about temporal experience, when we perceptually experience distant things, we are already possibly anosognosic of some of the relations of temporal order in which they stand.

Yet the external events also appear to be temporally ordered—e.g., I see the stars *then* go back inside the house and hear the conversation. Where that goes wrong, is it an illusion of temporal order? Could it be some kind of anosognosia—e.g., could this appearance of temporal order be the non-appearance of something more?

Here is my suggestion:

(1) We do not experience temporal order or duration between:

 (a) Perceptually experienced external events. For example, we do not see the actual order or duration of star shine in seeing the star shine.
 (b) External events and events that are part of our subjective activity.

(2) Some perceptually experienced external events appear to be simultaneous with each other and with different stages in our subjective activity.
(3) We do experience temporal order of subjective activity. That is, we do experience the temporal order of events between stages in our subjective activity.
(4) If one external event E appears simultaneous with one subjective stage A and another external event F appears simultaneous with another subjective stage B, then the temporal order of (A and B) appears to be the temporal order of (E and F).

Option (4) supports my suggested account of the alleged illusions of temporal order. However, to be clear, it does so by exploiting the complexities of the processing in experience between external events and the related subjective activity.

The account goes like this:

> Each stage of a perceptual experience begins at the external events (or distal stimuli) and ends in subjective activity. All the elements of that stage appear to be at the same time.

The order of each stage's start—determined by the stimuli—and the order of the stages themselves are independent of each other. The external world does not directly determine the order of stages. The external world directly determines the beginning of the stages.

Here is an analogy: A line of ropes mooring a boat lies along a quay. The same ropes also line the side of the boat they tie to the quay. The ropes cross one another on the way to the boat that they moor. As such, their order in the line along the boat is different to their order in the line along the quay.

Each stage can end before another begins. However, it can also end after another begins.

Next, we have this: Each stimulus or set of stimuli need not correspond to only one perceptual experience. Stimuli can be constituents of more than one perceptual experience; they can be overlapped by multiple experiences; experiences can share them (one might see this as inspired by but not dependent on Dainton's *overlap model* (Dainton 2000; Dainton 2010)). Using the earlier analogy, this is similar different ropes tied to the same place on the boat but tied in different places along the quay.

Thus, given several individuated experiences, the same stimuli can constitute each of the experiences yet in a different way. In one, the same set can be ordered A then B. In another, the order is B then A.

If we accept that, then the illusions noted earlier disappear. They are due to something else: an inability to be aware of the stimuli's actual temporal order.

However, there is a second part, and this is perhaps less comfortable. This is an inability to be aware of the order of those stimuli in other earlier and later experiences.

Yet this is also not an illusion. It is anosognosia, although it is a very specific kind. It is anosognosia of the relativity of relative properties.

9.3.5. The Relative and the Absolute

In his debate with Einstein, Bergson argued that relative simultaneity (especially simultaneity defined arbitrarily) conflicts with our experience of time. This is because our experience of simultaneity seems to be of an absolute simultaneity (Bergson *op.cit;* discussed in Power 2010). If relativistic physics is true, then this makes our experience conflict with reality, and so be erroneous.

We might question this. We might question the claim that our experience conflicts with the reality given relativistic physics. For what does it mean to experience absolute simultaneity? Does it mean a simultaneity that holds between all things, a universal simultaneity? Obviously not, as no one experiences all things. It means an experience of a simultaneity between some things, but which is *not* relativized to anything. What matters is that it does not *appear* to be relativized to anything.

In my 2010, I argue that we should conceive of the appearance of simultaneity as the non-appearance of temporal difference (e.g., duration, succession). If that is right, then the appearance of simultaneity where there is none is already not an illusion; it is a case of anosognosia. However, there is also another way there can be anosognosia here, one relevant to apparent passage (in a way that the earlier case is not).

I suggestion the following for relative simultaneity: Experience conflicts with reality by leaving out *that which defines the relativity of simultaneity*.

It is possible that the properties[7] that one perceives something as having are only had relative to a perspective or frame occupied by one's subjective activity—e.g., the place occupied by the subjective activity, the time in which the subjective activity occurs, and even (extending the range of subjective activity) the motion of the subject's body parts. From other perspectives or frames, the thing one perceives has different properties than those that one perceives them as having. However, this does not mean that one is having an erroneous experience.

It can appear as if those properties of the thing are relative in this way. For example, the visible parts of what an observer sees can appear relative to the observer. To use a common example in perceptual literature, a coin can appear to be tilted away from you, appearing to have certain (but not all) properties visible, and for what is visible of the coin to appear to have depth, height, and width. These depths, widths, and height can appear to be perspectival properties, that is, properties relatively determined by the subject's point of view.

This appearance of relativity can come through the perceptual experience itself, e.g., a shape appears to be tilted away from you in three dimensions, or through the complete experience, e.g., through amodal completion a shape appears to be three dimensional, much of the three dimensions being hidden or imperceptible.

However, one can be under an erroneous experience if the following occurs: The experience is of properties (or particular values of properties) that are *invariant* under all perspectives (or frames). For example, say that a certain shape, motion, or colour are frame-dependent (like the visible portions of the coin) yet appear to be perspective-less and invariant (like the three-dimensional shape of the coin under Newtonian physics). If they do appear perspective-less and invariant, then this is an error.

Given this error, these properties appear to be absolute. They appear to be the same no matter the perspective from which we they are observed. Yet they are not. They vary by perspective and are relative.

So is this erroneous experience of an absolute property an illusion, hallucination or anosognosia?

It is anosognosia. One is unaware that one's own perspective or frame defines the properties that one observes, or one fails to experience the limit of the perspective (e.g., how it is incompletely observed is occluded by something else, or otherwise appears as it does due to this particular perspective). As a result, one fails to include the perspective in the analysis of the observed properties. These particular properties (e.g., the shape, colour, or motion) are treated as being the same for all perspectives or frames—i.e., as they appear to the observer in this perspective.

We have this with the three-dimensional shape of a gold coin in a relativistic universe. Its particular three-dimensional shape depends on the frame. If accelerated to near-relativistic speeds along its diameter, the coin's length contracts until its faces are oval rather than circular.

Furthermore, one can be under this error even if one is aware that one has a perspective. This is because we can be aware that we occupy a frame or perspective, but it is not a *relevant* frame or perspective. We think the frame does not determine these (apparently absolute) properties. If they do not appear to be determined like this, then we believe this is true—even though it is actually false.

This is relevant to tenseless theory and the experience of temporal passage. We can re-interpret the apparent tense properties of temporal passage as an appearance of absolute properties that are actually the non-appearance of relative properties.

Relative Properties and Passage

According to tenseless theory, tenses are relative properties. 'The past' is the *earlier than* relation with an assumed index to a time. It is *past to* some tenselessly defined time *t*. Similarly, 'the present' and 'the future' are the *simultaneous with* and *later than* relations with an assumed index, giving *present to* and *future to* some tenselessly defined time t.

If you vary the tenseless time—e.g., from *t* to *t'*, then you vary the tense. For example, an event may be present at *t* but past at *t'*. We can use that to give a tenseless account of the perceptual experience of passage, one that involves anosognosia.

As already discussed, there is reason to hold that there are perceptual experiences of tenseless change. This is a perceptual experience of a change in something's non-temporal properties. We can also say it from the *t* of the first stage of the change to the *t'* of the second stage of the change.

If we experience any events earlier than that experienced change, then in experiencing a change from *t* to the later *t'*, we also experience a change in the relative pastness of any events earlier than the experience. From t', any past event is the duration between *t* and *t'* more past than from *t*. For example, if the duration between *t* and *t'* is one second, and a past event E is three seconds ago according to *t*, then E is four seconds ago according to *t'*.

Similarly, if we experience any events later than the perceived change, the variation from *t* to *t'* in the change gives a difference in the relative future of the future event. From *t'*, any future event is the duration between *t* and *t'* less future than from *t*.

This is not a variation in the tense of the *perceived change*. The whole of the perceived change is present (or, rather, it is tenseless, if an experienced present x is just an experienced x). The presentness of events does not vary in the tenselessly perceived change.

Still, if one wants all perceptual experience of change to be tensed change, then this account can accommodate that. By having a perceptual experience of something happening at *t* and then at *t'*, any perceptually experienced stage at *t* is present (from *t*) and then is past (from *t'*).

The main point is this: We are not aware of the relativity of what we experience, of these events changing tense because of the change in the events' relationships to what we perceptually experience. For example, our experience could be of a change:

- From a present event E that is some past duration D from an earlier event X
- To a later present event F that is another past duration D' from X

D' is larger than D, and so X is more past when it is D' than when it is D. In our experience of E then F, it appears that X also changes its pastness (by going from D to the longer D').

Lastly, given the tenseless account of the change, there is no change in the tense of what we perceptually experience. As such, the change does not appear to be a change due to a change in the indexing location of the experiencers (i.e., the *t* and *t'*).

Instead, this appears to be an absolute change in the tense of the past or future events that we experience—i.e., absolute temporal passage. That is, there is an erroneous experience of what is, in fact, relative being absolute. If the previous analysis is correct, then this is anosognosia of the relativity of what we experience.

Relative Properties and Temporal Order

At the end of the last section, I suggested that we do not have illusions of temporal order. Instead, we perceptually experience temporal order, but it is order in subjective activity itself. Then, along with experiencing such subjective order, we perceptually experience everything else (including the stimuli) as simultaneous with stages of that subjective order. In addition, we are not aware of the subjective order of those stimuli in earlier and later experiences.

This kind of error is anosognosia of that order. However, it is also something else. We are also not aware that there is a difference between our experiences of such order. Such temporal order does not seem relative like this. As such, it appears to be absolute. That is, this is anosognosia of both the order of other experiences and the relativity of the order in our own experiences.

Much of this comes from not being aware of the durational differences between perceptually experienced events and between ourselves at other times and the present (however that is defined). That is, this is a failure of or a limitation of our perceptual ability to detect differences in

- Temporal properties in the structure of each perceptual experience
- Between different perceptual experiences

I think that this failure or limitation can account for much of the error in our experience of time. However, that that can do so does not mean there *is not* such hidden temporal structure in our perceptual experiences of time. It means there *is* such structure.

In many ways, this account is like the account of the perceptual experience of distant things in light of a time lag. The difference is that the lag extends into the subjective activity, including between stages of the subjective activity of the experience. If we already accept it for the objective activity in the perceptual process, then I can see no obvious reason why we would deny it for the subjective activity. The difference between the objective and the subjective does not seem to require that the latter have no time lags. If one thinks it does, I expect some argument to support it (and I cannot think of any).

9.4. Models of Temporal Experience

One reason to deny that perceptual experience has such structure is because perceptual experience must be *at a time*, and 'at a time' means that a perceptual experience has all of its structure at a single, durationless moment. In that case, it must have static structure. It must be a state. It can represent differences in time—it can represent temporal structure—but it cannot have differences in time between elements of its own structure.

A motivation for this denial comes of presentism. Yet one might think that, putting presentism aside, there is still a reason to insist in this denial. This is because one has models of time perception that require a denial of temporal structure.

Models are a way in which one can examine how considerations about positions on time can affect the understanding of temporal experience's veridicality. To close this chapter, I consider the most discussed contemporary models of temporal experience.

As with experience in general, *models* of temporal experience—in particular, perceptual experiences of time—are models of the underlying structure that gives us the temporal experience. They do not concern the content as such; they concern the vehicle structure that explains why the content comes about.

For the perceptual experience of time, there are two broad models discussed in contemporary theory—both within philosophy and in other disciplines (e.g., cognitive science). These models are *retention theory* and *extensionalism*.

In this section, I discuss the differences between these models in light of the phenomenology and the philosophy of time.

9.4.1. Retention Theory

Retention theory or *retentionalism* is characterised as a description of the phenomenology of time that theorists generally accept. Theorists often cite is as being a description of the phenomenology of temporal experience, whatever the kind of experience (for example, Kelly *op.cit*, Grush *op.cit*, Varela 1999, and Gallagher (*op.cit.*)).

Yet it is not such an account. What it significantly provides to the discussion of temporal experience goes beyond the phenomenology of experience to its structure and constitution.

Retention theory has its roots in Husserl's work on the phenomenology of time (e.g., Kelly 2005). It is allegedly an account of the phenomenology or appearance of time. So conceived, retention theory seems to be an account of temporal experience in terms of its content. As such, we might think that retention theory is an account of temporal experience.

The general form of retention theory is tripartite. Along with phenomenology of present things (often referred to as *primary impressions*), it posits two further classes of phenomenological features: *retentions* and *protentions*. These three features (retention, primary impression, protention) provide the structure of any experience that is of things *over time*, including perceptual experiences of change. Briefly, 'retention' and 'protention' are

- *Retention* (or what Husserl also sometimes calls 'primary memory'): an intention-like[8] attitude directed towards or about the past, often explicitly restricted to the 'recent' or 'just' past. It is different to what is more commonly considered consciousness or intentionality towards the past—i.e., episodic memory or memory experience (which Husserl calls *secondary memory*).
- *Protention*: an intention-like attitude directly towards or about the future, often explicitly restricted to the 'near' future. It is different to what the commonly discussed consciousness or intentionality towards the future—i.e., anticipation or expectation.

With those concepts defined, we can give the following account of time consciousness: Experiences of time are of events changing by being future and then present, and becoming present and then past. In the model, the retentions capture or stand in some way for the past events. The retentions are about (they *intend*) the past events. However, the retentions are not themselves past.

How we get a sense of even further past events without memory is by there also being retentions of retentions. We experience a retention of one event which is just past and also experience along with it a retention of *a retention of another event*, which is even more past, and which we used to experience as being just past.

For example, there is a sequence of notes: *A then G then E*. I have a perceptual experience, a perceptual appearance of *A then G then E*.

Figure 9.1 illustrates the retention theorist's account of this.

When *G* is experienced, there is a retention of *A*. The retention is labelled *A'*. When *E* is experienced, there is a retention of *G* (labelled *G'*) and a retention of a retention of *A* (labelled *A"*).

There have been doubts that these elements correspond to the apparent features of temporal experience. For example, Dainton argues that we do not experience the extra element of past retentions (Dainton 2000; Dainton 2010). Kelly thinks Husserl's account is correct, but is puzzled about how it explains the perceptual experience of change (what he calls 'pace perceived'; Kelly 2005).

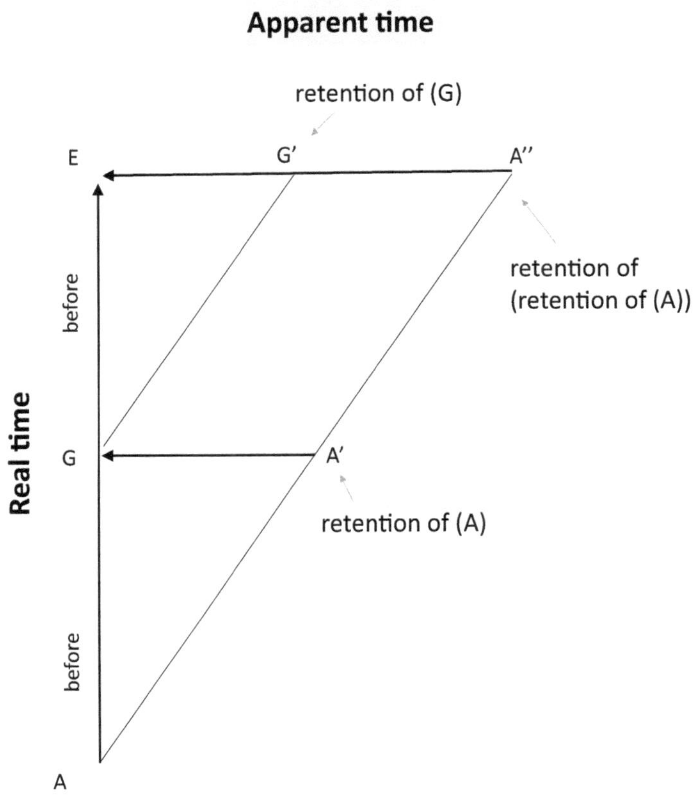

Figure 9.1 Retention theory

Given Dainton's views about time, my views can be can be seen as a rephrasing or as another way of approaching the issue from the same broad metaphysical direction. In any case, I do not think retention theory captures the perceptual experience of change in a phenomenological or naïve way. I do not even think it captures the tenseless or tense experience of change.

The most relevant concern to this discussion is that retention theory does not seem to be neutral to different metaphysical concepts of time. It seems to presuppose tense theory in how it raises problems and tries to solve them.

Retentions demand of the structure of temporal experience that there is a sub-period of that time from which the experienced times are divided into primary impression, retention, and protention. This makes sense with tense theory: The experience is in the present but is always of more than the present, and so needs a retention for what is earlier (the past) and a protention for what is later (the future).

For example, if we compare it to the tensed account in how the specious present of perceptual experience can be described in light of different metaphysical concepts of time, we see a common approach: an analysis of *any* experience of change (even changed which is perceptually experienced) in terms of the past, present, and future. This raises puzzles; for example, Kelly asks, how can we perceive the past or future or how can the present have duration (Kelly 2005)?

Yet, as discussed earlier, these puzzles do not occur with tenseless theory. We need not perceive either the past or future; the present can have duration.

If we are addressing Husserl's own analysis, and if tense theory and presentism are more intuitive than their alternatives, this is unsurprising. Husserl was not adopting counter-intuitive theories of time, or working explicitly on the metaphysics of time. He was carrying out a phenomenological reduction of time consciousness (Husserl 1991; De Warren 2009).

Yet once we move away to an account of actual perceptual experiences of time (and away from simply adapting Husserl's views on them), retention theory is not as obvious as a description of the experience. If this account is about what explains the experience, then, given tenseless theory, the retentional (and protential) structure looks redundant. We do not need to divide a perceptual experience of tenseless time into parts directed at the present, past, and future. There is no need for such distinctions in tenseless theory. For example, a tenseless account of the perceptual experience of change need not involve an experience over the past and/or future, and so need not involve anything directed towards the past (such as retentions) and/or future (such as the past).

As such, if one assumes tenseless theory (or even eternalism in general), then there does not seem to be a need for anything which one might call 'retention'. 'Retention' implies something retained from an earlier time. However, why retain something from one time to another to explain our experience of it? The only forceful reasons I can think of are as follows:

(a) One is a presentist who holds that we must explain anything we experience by using something real in the structure of the experience. As such, we cannot explain the experience of the past by using the past, because it is not real. We need something present—such as a retention—to explain it instead.
(b) One is a tense theorist who holds that we must explain what we experience by something present in the structure of the experience. If so, this makes it the same as presentism (one might call this *phenomenal presentism*).
(c) One thinks that we must explain what we experience by something strictly at one time in the structure of the experience. That is, we explain the content of experience by a static structure of an experience. It is not enough that there is dynamic structure of experience.
(d) There is good empirical evidence that only static structure in experience correlates with any experiences of time.

Options (a) and (b) involve commitment to theories of time that are not tenseless theory.

I do not know of any reason or evidence of (c) and (d). Furthermore, if my arguments from relativistic physics go through, there is, strictly speaking, no such evidence. Elements arbitrarily form static structure. If the structure that explains experience is not arbitrary, then it is not a static structure.[9] I assume that models of temporal experience such as retentional models do not posit arbitrary structure.

One may still pick out a special time for the perceptual experience of time, one from which the tripartite structure does work. My point here, however, if it is unnecessary, why do so?

In reply, a defender of retention theory might press the claim that we are motivated to retention theory because it corresponds to our phenomenology—it is how things appear to us (e.g., Almäng 2014). However, as said, there are theorists who do not think that this is so. For my part, I do not think so because, as discussed, I do not think analysis in terms of past, present, and future sufficiently exhausts the phenomenology of change. Change over days or weeks or months might be analysable in terms of those tenses, but *perceptually experienced* change is not.

However, even if all experience of change was of tensed change, retentions (and protentions) do not capture that change. The elements of the experience that are retentions (and protentions) are not part of the phenomenology. We are not aware of them as retentions and protentions.

Whatever it is that one experiences of the past, it does not seem to be a *retention* of the past. A retention is something held on to or kept present.

What is past is not the same as what is held on to. It is not present. A past thing is not a retention (and a future thing is not a protention).

That does not mean that something present, such as a retention, cannot explain the phenomenology of something past. It can stand for, or represent, or be directed towards the past. However, if a past thing is apparent to us, its stand-in (or representation) is still not the same as that appearance.

That is, retention may explain a phenomenology of pastness, but it does not do so naïvely. It does so by either denying how things appear is how things are or by ignoring how things appear. Neither kind of denial or neglect is part of a phenomenological account of temporal experience. It is also not a naïve account of temporal experience, realist or otherwise. It is at best a part of the possible hidden structure of experience, such as subjective activity, imagination, or representation. Non-naïve perceptual theories may include such an account, but that does not mean that naïve theories can include it (the theories I am interested here).

Does this mean that retention theory requires some kind of erroneous experience? Yes; however, the kind of error depends on what sort of role it plays in its relationship to the experience of past things.

First, a retention may be an additional element in the experience of past things, playing some role in the structure that explains our awareness of them, but is not otherwise sufficient for that part of the experience. In that case, retentions might be something hidden from the subject of the experience. If so, then we are anosognosic of their presence in the experience. However, we also experience past things as well, and the past things are a separate part of our experience.

Second, the retention alone gives us our experience of past things. No a part of the experience is a past thing that is separate to the retention. The part that is the experience of the past thing is identical to the retention. In that case, retentions involved an *illusion*. Whatever gives our experience of past things (the retention) only appears to be identical to those past things (the past is not the retention).

In that case, depending on what is meant by 'past things', we may either have illusions or hallucinations with retention. We have illusions if what is past is a property and hallucinations if it is an object. In addition to those errors, we also have errors of tense, of course. The retention is present; what is 'retained' (except they're not, being past) are not present.

It is worth noting that if we perceptually experience tenseless change, then retentions can at least be the same tense as what they are about. However, being retentions, they are still at a different time. The stage of the change retained is earlier than its retention. The analysis for tense theory is also applicable to tenseless theory. The only difference is that instead of a difference in tense, there is a tenseless difference in temporal location.[10]

Retention theory may still seem necessary in some way to you. If you still think so, then I ask you to check that your thought is not based on presentist or tense assumptions. From my perspective, with those put aside, I do not find it necessary.

9.4.2. Extensionalism

Another account of the *phenomenology* of time and the constitution or metaphysics of temporal experience is *extensionalism*.

Dainton coins the term extensionalism (Dainton 2000). It is the position that our experience of time (or, better, of temporal extent) comes from our experience itself extending in time. The actual constitution of the experience explains the content. How things seems—a temporal extent—is how things are: The perceptual experience *is* temporally extended.

For example, I hear a sequence of notes: *A then G then E*. Figure 9.2 illustrates the extensionalist's account of this.[11]

At least three elements constitute my perceptual experience. There is an element of the experience that is (in some way related to) A, an element that is related to G, and an element that is related to E. These elements are temporally ordered and they exist over a duration. This order and duration are apparent to me. This temporal order and durations gives the perceptual appearance of the sequence and duration of AGE.

This is a naïve theory of temporal experience: There is an appearance of temporal features (order and duration); these features appear to be possessed what is experienced, and this is due to there being such temporal features in one's experience.

In addition, note that this account is compatible with my account of experienced temporal order earlier in the chapter. The order need not be the actual order of the notes; it need only be of the experience's subjective activity. The temporal order of the objectivity activity (e.g., the pressing

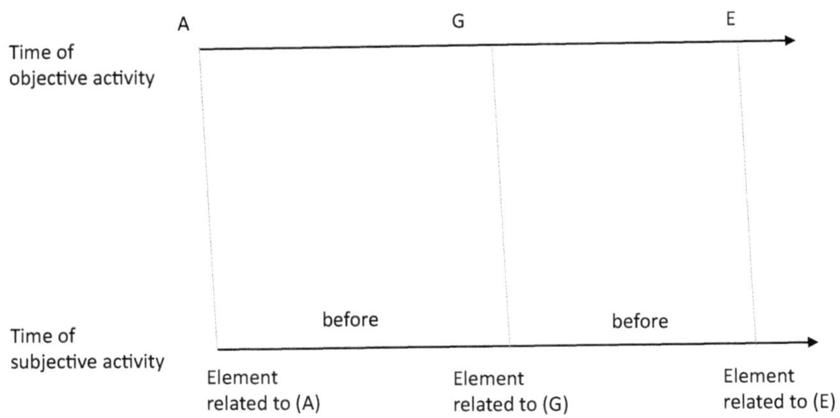

Figure 9.2 Extensionalism

of piano keys when playing the notes) need not be apparent, just as it need not be apparent when I look at the night sky.

However, depending on how things are, there is still a question of whether or not this is plausible. In keeping with the thesis of this book, this depends on one's philosophy of time. Take, for example, the experience of present change.

Tenseless theory: Several moments in a duration can be present (or indeed, past or future). It only depends on if the moments fall within the period of time that picks out the present (or past, or future). If extensionalism is true, we perceptually experience a series of events, or a period of time, by our experience itself extending over that series, or that period of time.

We already discussed what sort of erroneous perceptual experience this might be. If we are referring to experiences of tenseless duration, such as that of tenseless change, then there is no erroneous perceptual experience here. It is veridical.

However, if we are referring to experiences of tensed duration, such as that of a tensed change, then there is erroneous perceptual experience. Either there is an illusion of tensed change or duration, or anosognosia of the relativity of the tenses involved in the change.

Tense theory: Only one moment is present in a duration; all remaining moments are past (and/or future). If extensionalism is true, then this makes our experience of a series of events, or a period of time, due to the experience extending into the past (and/or future). That is, past events (and/or future events) constitute the experience. Furthermore, unlike retention theory, the experience is not only about past or future events; these events are constituents.

We already discussed what sort of erroneous perceptual experience this might be. If we are referring to experiences of tensed duration, such as that of tenseless change, then there is no erroneous perceptual experience here. It is veridical.

However, if we are referring to experiences of tenseless duration, such as that of a tensed change, then there is erroneous perceptual experience. There is anosognosia of the real tense change behind the apparent tenseless change.

Presentism: Only one moment is real; all other moments are unreal. As this is extensionalism, past events (and/or future events) constitute our experience. However, past and future events are unreal. As such, unreal events constitute our experience.

Recall that extensionalism means we experience a duration by the experience extending over the duration. Thus we do not just have an experience that represents, intends, or is about unreal events. It is *made* of them. That is, given extensionalism and presentism, we have a constitution or structure that is of unreal things.

What sort of erroneous perceptual experience might this be? If we are referring to experiences of tenseless duration, such as that of a tensed change, then there is erroneous perceptual experience here. There is an illusion or hallucination of all but one of the stages, for they are all unreal except for the present one—yet they seem to be real. If there is also tensed change in the present, we may experience it, and we can understand the error in our experience as anosognosia of that change.

Whatever our metaphysical presuppositions, these results are significantly different. Tenseless theory works well with extensionalism. Tense theory might work with extensionalism. Tense theory requires that part of something be past and future in the robust sense of those tenses insisted upon by tense theory. However, as already discussed earlier regarding the perceptual experience of change, if we hold that we perceptually experience temporal passage, and we are naïve about that perceptual experience, then we already have experiences that are constituted by past and future things.

However, presentism generates the implausible scenario that our experience is constituted by or extends over unreal things. This is more difficult to accept.

Thus, if tenseless theory is true, for the perceptual experience of time, it is not obvious that retention theory has an advantage over extensionalism. If tense theory is true, retention theory has experiences in the present that are about other times. It does not extend over those times, but only needs to be directed towards them (via retentions and protentions). This seems necessary given presentism, for there is nothing of those times to extend over. I do not know what eternalist tense theorists might otherwise think about this. I leave that for them to work out.

In this book, I assume that tenseless theory is right. Thus I prefer extensionalism and not retention theory. However, another hypothesis of this book is that many people think intuitively in terms of tense theory and presentism. If that is right, then this may explain the appeal of retention theory.

9.5. Conclusion

In this chapter, we looked at temporal experience, its errors, and some of its models. The purpose is to discuss why one might talk about experiences that are not just perceptions and to illustrate how thinking about errors of experience can feed into evaluating different concepts of time.

The issue with errors in the perception of depth, the experience of tense, and the experience of simultaneity and duration suggests that, even if different theories must deny such appearances, we may still evaluate them by how much must be denied about the appearances.

To summarise,

(1) *Temporal experience* is of something temporal. It is not just any kind of experience or perception, despite what we experience or perceive typically appearing to be in time.

(2) We can have erroneous temporal experiences. We can have temporal illusion and temporal anosognosia. However, it is implausible to suppose there is anything we might call temporal hallucination.
(3) The most notable kinds of erroneous temporal experiences concern tense, change, and simultaneity.
(4) We can evaluate models of time consciousness by the errors they commit us to. However, part of that evaluation requires weighing them in light of different positions about time.

This discussion does not exhaust all cases of temporal experience, or ways in which these experience can go wrong. I only focus on perceptual experience. I do not look at memory experience, anticipatory experience, or even a more extended experience of temporal passage, for example, one which includes non-perceived past and future events.

I leave that kind of work to the future. What I hope is given here are a few ways of thinking about the issues. In any case, there is enough to work on here with respect to perceptual experience.

A central desideratum of this book is to preserve the position that how things appear corresponds to how things are. This is so even where in its role towards a more complete knowledge of the world experience leads to false beliefs. With that in mind, we have been examining ways of describing and explaining such an appearance given different ways experiences can fail. As discussed, a theory committed to anosognosia seems best; a theory committed to illusion is the next best; a theory committed to hallucination is the worst.

This suggests another desideratum of a project around naïve realism and one which the philosophy of time has some relevance. This is the elimination of hallucination, which is the subject of the final chapter.

Notes

1. This difference rests on the assumption of a difference that some contemporary philosophers might reject. If one conceives of perceptual experience as including cognitive penetration then, given cognition is intentional, perceptual experience is intentional (or, at least, is partially intentional, the part determined by the penetrating cognition).

   However, this denial does not favour naïve realism. There is a phenomenological difference between perception and other cognition (such as thought). For discussion of illusions and cognitive penetration, see, e.g., McCauley and Henrich (2006).
2. Perhaps tenseless theorists might be able to hold this. The past cannot be partially constituted by the present given the same defining index—i.e., the same indexing tenseless time, but all there is to these tenses are their indexical character. So an event might be past to one index and present to another (as the counter in this cafe is here indexed to the cafe, but there indexed only to my table; see Power 2016). However, tenseless theorists do not have to do this.
3. Again, this is from a naïve standpoint about temporal experience. It concerns experiences and appearances. This does not mean that one must accept these

appearances; one can adopt a non-naïve position. For example, because of causation or the open future, one might deny that future events constitute one's experience. However, at this point of the discussion, such commitments are not relevant. They are only relevant if they come directly from the appearances.

4. Some philosophers may understand this as the claim that we have a 'feeling of succession', a feeling that is not itself a succession of feelings (e.g., Hoerl 2013 for discussion).

   Putting aside grammatical reasons for why that is not so (a feeling (singular) is not the same as feelings (plural)), I take it that the claim here is we ought not to understand the temporal order (the succession) of experience (the feeling) as simply giving the experience of temporal order. In some cases, I think this is what we can do and something we would want to do.
5. Paul uses this example to argue that the brain takes both stimuli to generate a representation of the multiple flashes, adding a representation of 'flow' to the representations of the flashes to give continuous motion (Paul 2010).

   I do not think that addition is necessary. I do not know how strong the flow is, or how perceptually experienced it is separate to the rapid sequence of the actual stimuli themselves. In any case, one can extend what I say here about colour to flow.
6. Technically, we could also get a case in which both are wrong, but the examples do not allow it. If that case, events cannot be either temporally ordered or simultaneous. This is to hold that there is no temporal order (some tenseless theorists may hold this) or that the events do not exist at all in time. I leave that for investigation beyond this work.
7. Alternatively, if one wishes to make such a distinction, properties and/or different property-values—e.g., the property is *spatial extension*—something has that property independent of the frame; however, the amount (i.e., value) of the extension varies from frame to frame. If one makes this distinction, then read all references to 'properties' as 'properties and/or property-values'.
8. Although, according to Gallagher, Husserl did not consider it quite right to think of it as an intention (see Gallagher 2003).
9. It may appear to be a static structure; however, that is a separate issue. That would make the experience more like the experience of apparently simultaneous spatially related things (see the chapters on spatial experience).
10. For example, Grush (2005).
11. This figure for extensionalism is not identical to Dainton's. I adapt it—and my description of the position—to suit the model of perceptual experience I advance in this book; in particular, elements, subjective activity, and objective activity are particular to my presentation; they are not found in Dainton's (or others') descriptions of the position.

10 Eliminating Hallucination

This last chapter is a move into more speculative territory. I present an argument against hallucinations that many might consider strange or counter-intuitive. I base the argument on an explicit assumption of tenseless eternalism—in particular, the reality of multiple times and the extension of experience over such times. The speculation in this chapter is an application of positions defended throughout this book. It considers how experiences might or might not satisfy the conditions of hallucination given tenseless theories of time.

I argue that, given tenseless eternalism there are ontological resources available to conceive of experiences that are *candidate hallucinations* (as I will call them) as something else instead: other kinds of erroneous perceptual experience extended in time. This is not an argument for the elimination of error, but it is an argument for the elimination of hallucination.

I have already consider one way to eliminate—or, at least, reduce—hallucination. I discuss this in Chapters 4 and 5 on erroneous experience. This is to take evidence of hallucination as only being evidence of a complex illusion. A complex illusion is a discrepancy between the appearance of a multiple of properties and the reality of those properties. The appearance of some multiples of properties is arguably the same as the appearance of an object. Given this similarity, we can provide the following analysis of evidence for hallucination: Such evidence for a hallucination of an object is, in fact, only evidence of a complex illusion of properties.

This may work to eliminate actual cases of hallucination. Using this analysis, whatever actual evidence there seems to be for hallucinations is indistinguishable from evidence for complex illusions. If such complex illusions are insufficient for hallucinations (i.e., objects are not just bundles of properties), then evidence for such illusions is not evidence for hallucinations. If that is the only actual evidence we have, then we have no evidence of hallucinations.

This method of eliminating hallucination is central to my approach in this chapter. For many experiences that we might interpret as hallucinatory, I suggest a tenseless (and eternalist) non-hallucinatory alternative interpretation. Part of my approach includes alternatively conceiving of hallucinations as complex illusions—i.e., as illusory experiences of 'objects' that are only illusory and conflated bundles of properties.

However, one may object to this attempted elimination of hallucination by asserting that there is, indeed, a great deal of evidence for compelling cases of hallucinations and hallucinations that are not just complex illusions of properties. There are cases of perceptually experiencing something when there is *nothing there*. For example, in flotation tanks and in dreams we experience objects with multiple properties. In experiences such as these, there is not just a lack of properties. There is a lack of objects. Experiences in flotation tanks and dreams are common and typical cases of hallucinations. Whatever our thinking about time, how we can eliminate such typical case of hallucination?

10.1. Evidence of Hallucination

When one claims that a subject is hallucinating, there usually seem to be good reasons to make this claim, no matter one's view on time. The subject experiences what seems to be something heard, seen, smelt, or even felt where, in actuality, there is nothing *there*. Alternatively, the subject experiences what seems to be something heard, seen, smelt, or even felt where, in actuality, there *is* something there. However, due to the structure of perception, they cannot perceive that thing.

There seems to be a great deal of evidence for both of these situations. People claim to see angels and other supernatural beings (such as the religious visions of my childhood). People describe hearing music where there is no music playing (e.g., Sacks 2008). They detect phantom smells. They feel the shape and pain of phantom limbs. They report these in many different contexts: suffering sickness such as fever or schizophrenia, under the influence of drugs, tired, suffering 'flashbacks' or memory hallucinations in post-traumatic stress disorder (PTSD).

People also perceptually experience colours and shapes where nothing can be seen of such colours or shapes—e.g., after-images, hypnogogic images (when subjects are falling asleep) and hypnopompic images (when subjects are waking up).[1]

Therefore, these are situations where we hold that a subject hallucinates: There are no angels; there is no music; there is no detectable smell; there is no longer a limb; there is no repetition of a traumatic event; there is no coloured shape; there are no things in the room when someone is waking or falling asleep.

Thus, with all of these examples on hand, how could we eliminate hallucination? How could we replace it with illusion, anosognosia, or veridical experience?

I assume that we cannot eliminate erroneous experiences. We do not have veridical experiences in these cases. I assume in these cases there are no angels, music, smells, limbs, traumatic events (except the experience itself, of course), colours, or other apparent properties or objects.

However, does this mean there must there be hallucinations—i.e., erroneous experiences of objects?

In Chapter 5, I argue that hallucination is the least desired form of erroneous experience for a theory. As such, above other kinds of error, it would be good if it could be relegated to a mere possibility, with all actual candidates for it comprehensible in other ways—e.g., as anosognosia or illusion. In the absence of such an ideal situation, it is still better to be able to minimise the number of cases of hallucination.

Yet to do this for any experience that is a candidate hallucination, we need to remove the one thing that distinguishes hallucination from either illusions or anosognosia: We have no veridical experience of an object. We can still have other errors of properties and relations. Yet, in the cases given earlier, how can we have it that there is a veridically experienced object?

One answer is that we have a veridical experience of some kind of mental or internal object—e.g., a sense datum. However, there is a central problem with this approach: A great deal of what people report hallucinating does not appear to be mental or internal. It appears to be mind-independent and outside of their mind or body. Thus, appealing to sense data requires that the experience be of something significantly different to what is apparent. It lacks externality and mind-independence.

As evident in the discussion so far, both externality and mind-independence are almost ubiquitous across all modes of perceptual experience. Only proprioperception, the haptic sense, or taste are exceptions, and even those of mind-independent things (assuming one's own limbs, physical surfaces, and food are not mind-dependent in a relevant sense). As such, a sense-datum analysis involves denying the perceptual veridicality of a fundamental property. That might bother some theorists, who might feel it is so important that we should try to preserve externality where we can. Therefore, it might motivate one to pursue another way of eliminating hallucination.

To satisfy the externality, we must take a different approach. This is where the philosophy of time is relevant. The sort of phenomena claimed to be cases of hallucination can be analysed differently given different concepts of time. Different concepts of time give different answers as to whether or not some object is there, whether we can veridically experience it—e.g., perceive it, and not just experience it. If we can perceive the object, then we do not hallucinate. Either we have anosognosia (of non-apparent things) or we have illusion (of apparent properties and/or relations).

That is, in order to eliminate interpretation of experiences as being hallucinations, there must be an object, and we must be able to experience it veridically. For example, when angels are visually apparent, angels are not seen (we can only see what is actually there). However, one sees *something*. Similarly, one experiences a piece of music now; one does not hear that music now but one hears *something*. One experiences a particular smell, colour, motion, or shape in a room; one does not smell that smell or see that colour, motion, or shape in the room, but one smells or sees *something*. One seems to feel the shape and pain of one's arm; one does not feel the shape and pain of one's arm, but one feels *something*.

So how do we eliminate hallucinations for the given examples? To answer this, we have two constraints:

(1) We must provide an analysis or explanation of experience without committing ourselves to what is central to hallucinations: that there appears to be an object, yet there is no object. Candidate hallucinations must turn out to be veridical experiences of objects.
(2) As there are no angels, ghosts, etc., we cannot avoid hallucination by insisting that experiences of the same are veridical experiences. Candidate hallucinations cannot be veridical experiences of everything there is about an object. There must be an error of some kind—e.g., of the properties or relations of the object.

To satisfy both these conditions, consider some answers which do not involve any particular commitment to a philosophical position about time.

10.2. Imagination

The following approaches to hallucination may seem to explain, or even eliminate or reduce, hallucinations by appeal to something else. These are explanations of hallucinations in terms of imagination, internal objects, and memory experience. However, given the core aspect of hallucinations, and the motivations for explaining them, these approaches have problems. Their main problem is that each approach either cannot eliminate a central feature of hallucinations: that there is an appearance of an object yet there is not an object or else they deny a fundamental property of the object, that it is external or mind-independent.

In Chapter 3, I discuss illusions of imagining, experience that appear to the subject to imagining but are not. For many cases of hallucination, what is hallucinated appears to be real. It does not appear to be imaginary. Could this be all there is to hallucination? Hallucination is a mistaken form of imagining, and this eliminates hallucination? For example, someone who reports perceiving something that they cannot perceive does still *experience* the limb. They have an experience that is a sensory *imagining* of the limb. There is no perceptible limb or object. However, unlike the case for illusion, anosognosia, or veridical perception, there need be no object when one is imagining.

This does not make much progress towards resolving the challenge of hallucination, and especially not towards eliminating hallucination. Indeed, it is more a compounding of the problem of hallucination. It is a re-description of hallucination, with one extra condition: The subject is also mistaken about what sort of experience they are having. They seem to be having the experience had in perception, but they are only having the experience had in imagining.

This is a problem particular to imagination's use in explaining hallucination, rather than in explaining other kinds of experience. To see this, consider obvious imaginative experience—where it seems we are imagining—and amodal completion—where it seems we are aware of something hidden or absent. In both cases, what is experienced is in some way perceptually absent. This looks to involve some kind of experience of absence. In contrast, the experience in hallucination does not appear to involve imagination.[2] It appears to be a perceptual experience—of something real and present, not unreal, hidden, or absent.

As such, we need to take a significant step to bring hallucination under imagination, more than, for example, what we need for amodal completion. Something experienced that is absent, hidden or unreal has to appear sometimes as present, obvious, and real. This difference in appearances is significant.

There is another issue with imagination. With respect to what is most significant about hallucination as an error, imagination is the same as hallucination. Explaining hallucination in terms of imagination is not an elimination of hallucination.

Hallucinations are experiences in which there perceptually appears to be something that we do not perceive because it is unreal or we cannot perceive it. A condition of perception is that what we perceive is real. The appeal to imagination moves us explicitly to the first: There seems to be something that cannot be perceived because it is unreal. Appealing to imagination does not change that, and it is that which is central to the distinction between hallucination and other errors.

Imagination is not an elimination of hallucination; I hesitate to call it even an explanation. It is an account that fails to capture the phenomenology and does not remove the most problematic aspect of hallucination.

The point is that, to eliminate hallucination, one needs to have the object real and perceptible. To satisfy this phenomenology, one needs to capture the reality of what is experienced. This leads to the next approach: an appeal to memory experience.

10.3. Memory Experience

Perhaps hallucinatory experience is an erroneous form of memory experience. We have an episodic memory, but we do not know it is an episodic memory.

This has some advantage over imagination because unlike imagined things, the ontological status of remembered things is that they might be real. They are past and (depending on what is true of time) past things can be real.

If presentism is true, past things are not real. In that case, the memory account loses that advantage over accounts in terms of imagination. If we seem to perceive it, as in hallucination, we are in the same position as we are with imagining: What we perceive seems mistakenly to have existence but is, at best, something intentional and inexistent (of course, given time lag, we are in the same position with perceptual experience in general).

In contrast, if eternalism is true, then what we remember is real. It is only at a different time. This opens the possibility that we remember something in a similar way to perceptual experience. We experience something real. It is only that we do not perceive it. Instead, we remember it.

However, despite the past's reality under eternalism, it is not enough to eliminate hallucination. A hallucinated object can be real if we are unable to perceive it. In naïve realist terms, to avoid hallucination, the perceiving subject needs to be perceptually related to the object and/or the object needs to be a constituent of the perceptual experience.

Given eternalism, can memory experience such relational and constitutional conditions? There are reasons to resist saying that they can.

The phenomenology of a veridical memory is not the same as the phenomenology of perception. Where it appears to be a memory, a memory experience appears to be of past things, which is different to perceptual experience. Candidate hallucinations are (erroneous) perceptual experiences. They do not appear to be of past things (at least, for the examples I discuss here). Hallucinated objects are present, just as perceived objects are present.

This might be satisfactory. Sometimes memories seem to be of present things, and thus seem to be perceptions. However, this account of failed memory experience also covers perceptual experience in general. Recall the earlier discussions on time lag and tense. With a tense theory of time, and time lag, our perception is of past things that seem to be present. This makes it the same as this memory account of hallucination.

The time-lagged perceptual experience is a perception because it is of external things. The external things are past. However, it is a perceptual experience because it appears to be of present external things.

A hallucination-as-memory experience is a memory experience because it is really of past things. However, it is a perceptual experience—one that makes us think it is a hallucination—because it appears to be of present things.

So what could be the difference between memory and perception? The only difference seems to be a condition on memory that is missing from perception and is missing from the phenomenology in candidate hallucinations. This condition is both phenomenological and ontological. It is about how memory appears to be constituted and related to the world, and how memory is constituted or related to the world. It also gives reasons to resist holding that past things constitute memory experiences—and so that we can eliminate hallucinations by appeal to memory experiences.

In perceptual experience, we seem to be related to external real things that are constituents of our experience (indeed, given what is presented in the experience, the only constituents). As such, sense-datum theory or even contemporary forms of representationalism are problematic for perceptual experience. They undermine the appearances in their basic conception: Sense-datum theory puts something extra between the external things and the subject; representationalism claims representations can generate the appearance of external things without the need of external things.

However, indirect realism or representationalism do not seem so problematic for the phenomenology memory. According to Le Poidevin (2007), the claim that we directly experience past events in memory is not as intuitive as the assertion that we directly perceive events. That remembered events are constituents of memory is not so obviously apparent to the remembering subject.

As seen in the earlier quote from Husserl (Chapter 3), descriptions of memory experience by phenomenologists do frequently refer to what is apparent in memory as being some past event. Yet it is also seems that memory involves something intermediary between the subject and the past event (see Locke 1971; Le Poidevin 2007). In some way, it *seems* to me that in having a memory experience of some object O, the following holds:

(1) I had a perceptual experience of O.
(2) The perceptual experience of O is earlier than the memory experience of O.
(3) The perceptual experience of O plays a *mediating* role between O and the memory experience of O.

Given this analysis, we might get candidate hallucinations from memory experience as follows:

(a) If a memory experience of O does not include *having had a perceptual experience of* O, then the memory experience of O seems to be a perceptual experience of O.
(b) We have a memory experience of O.
(c) We had a perceptual experience of O.
(d) In having the memory experience of O, we do not experience it as including *having had a perceptual experience of* O.

Thus we have a memory experience of O that seems to be a perceptual experience of O.

Lastly, we get candidate hallucinations by having a perceptual experience of O that is of a remembered O. O is as real object (given eternalism); as long as this experience has O as its constituent, this experience is not a hallucination.

We might also remember that we had a perceptual experience of O. Yet, in that case, this erroneous perceptual experience of O appears to be of O happening *again*. What is missing from that account is that this experience contains the previous experience as part of its content.

For example, PTSD sufferers may have a memory experience of a traumatic event that is not like remembering a past traumatic event but is like a perception of the event. In that case, why it does not seem like a memory is because the experience does not include having had a perceptual experience of the traumatic event. They may also remember that they previously experienced it and so that this experience is a repeat of that earlier one rather than a remembering of it. However, this additional remembering is separate to the flawed memory experience.

Thus, hallucination is a case of the following going wrong: a memory experience that has gone wrong because it does not seem to be mediated by a previous experience. Instead, it seems itself to be that experience.

There are two problems with this account:

First, the account so far does not work with hallucinations of things and events that never happened, even in the past, such as angelic visitations. At the very least, we need a more complex account. This is a disadvantage it has in comparison to analysis in terms of imagination. In those cases, we just have to deny there are any such objects. I return to this problem in the next section, when we address it with the delayed perception account. What applies there also applies here.

Second, the aim here is to eliminate hallucination while accepting other kinds of errors. Memory experience here has gone wrong, but how it goes wrong seems to raise already discussed problems with previous attempts to eliminate hallucination? If we accept one account, what I shall call the *diminished-perception* account of memory, memory avoids hallucination but involves a demand upon the structure of memory that, in cases of veridical memories, may seem too much. If we accept another account, what I shall (and many others do) call an *image* account of memory, we preserve an intuitive account of memory. However, where they occur, we do not seem able to eliminate the main component of hallucination, that there is no experience of the object.

(a) Diminished-Perception Memory

In this account, we distinguish memories from perceptions by memories being less vivid or phenomenologically less rich or compelling. However, this is not due to the phenomenology or contents of the memory being in some categorical way different.

The phenomenology would be the same as perception except for the following in any memory (veridical or otherwise): Memory includes a distancing *mediation* between the subject and the contents. A remembered thing does not seem to be experienced directly, but through something else. This is the 'having had a previous experience' of the memory content. Being through something else, the experience is weakened in the vivacity, richness (etc.). Here is an analogy: In seeing something through a murky pool, it looks less vivid or detailed than seeing it directly. The same is the case with veridical memory experience: It seems to be experienced through something else and thus to be weaker.

In this account, hallucinations occur when this distancing mediation is itself not apparent, is hidden, transparent, or otherwise undetectable. It is still there—hence, this is a memory—but we do not know it through the experience. As such, what is experienced seems as fully vivid, rich, or compelling as perceptual experience. To use the same analogue, if the pool were perfectly reflective, what is seen looks as vivid and directly seen as what is seen without the mediation of the pool.

What this account does is demand that the candidate hallucination is just like a perceptual experience, except that there is some indirectness to the experience that is not apparent to the subject.

The problem with this account is that it means that, when we have veridical memory experiences, all we have that distinguishes it from perceptual experience is that the mediating feature interferes with the vivacity of our experience.

We might even argue that the memory experience is inaccurate, and the hallucination accurate. After all, what is experienced in the hallucination is real. It is just at a different time. However, as I discuss in Chapter 6, tenseless eternalism means that this difference in time does not mean that it does not have the same properties as any things at the time of the subjective activity. In this account, we may argue that some significant structure is hidden from us—i.e., we are anosognosic of it. The structure about which we are anosognosic is the mediation by a previous perceptual experience of our current experience.

Yet if we have no other reason to think so, why even think that there is such hidden mediation going on? Perhaps we are having an unmediated experience of something long past? Again, that we previously experienced something may be evidence of it. That does not mean that we are experiencing it now through that previous experience. After all, that is not how it seems. If we are not experiencing it in such a mediated way, perhaps we are not have a flawed memory experience at all.

(b) Image Memory

Some discussions of memory suggest that memory seems to be directly of past things (e.g., the quote from Husserl in Chapter 3). It does not seem to be of a present thing, through which we experience past things. Still, that does not seem exactly right. As discussed earlier, when we have a memory experience of E, we also have some sense that we experienced E before.

A natural account is that we do so by experiencing a present image of some x that we interpret as being the past x. Thus, like the previous account, there is mediation. However, there is a difference: In this image account, we do not experience one thing *through* another thing, but one thing *as* another thing.

We then get hallucination as follows: we experience a present thing not as a past thing, but as another present thing. For example, we experience a present image of a horse as a present horse. In this way, we hallucinate.

What distinguishes this image account from the previous account is that this involves an experience of an internal thing. The previous account does not require that—it depends on how one understands the mediation ('through' is not necessarily identical to 'as').

In addition, our normal memory experience is illusory, involving internal experiences: What seems to be past and external is, in fact, present and internal. It just seems to be past and external.

As such, we have a similar result to the intermediary account: hallucinations seem to be more veridical than the memory experience to which they are related. With the memory experience, we experience one present thing as being a past thing. Here we fail to experience a present thing as past because we experience it as present. However, this is not an error. It is accurate: We are experiencing a present thing as present. This case (when we are hallucinating) misleads us less than the normal case (when we appear to be remembering).

In an eternalist model of the structure of experience, the analysis of hallucination in terms of memory seems to involve either a hidden mediation or an internal thing. We are reducing hallucinations to other things. In these cases, respectively, to anosognosia of mediating experiences and to illusions of internal things. If we take a presentist model, of course, we are appealing to either non-existent intentional things (the experienced past things) or internal things (the images of past things).

As such, the appealing to memory in many cases is an appeal to intentional entities or representations. The exception is the mediating memory experience, to which I return later in this chapter.

As a final point on both imagination and memory, one might wonder what the advantage is supposed to be in positing intermediate or imagined objects. Why not keep hallucinated objects? Neither deals with the appearance of the mind-independence or externality. Both intermediate and imagined objects are mind-dependent in some way.

So what, in such a case, are better alternatives to hallucination, mistaken imagination, or mistaken memory experience?

10.4. Eternalism and Hallucination

In this chapter, I consider the case of hallucinations and time from the side to which I am committed: tenseless eternalism motivated by relativistic physics. Here is a reminder of some consequences of those positions:

(1) *Eternalism*: Anything at any time is real, irrespective of which time, for whatever reason, is past, present, or future.
(2) *Tenseless theory*: The past, present, and future are defined relative to temporal points of the B-series, itself defined by the relations of simultaneity and succession.
(3) *Relativistic physics*: The relations of simultaneity (and some relations of succession) between spatially separated things is arbitrary.

Furthermore, I assume that perceptual experience is

(4) Spatially located and extended;
(5) Constituted by elements extending through space and located at different places to other elements;
(6) Constituted by elements extending through time and located at different places to other elements;
(7) Constituted by *subjective activity*; and
(8) Possibly partially constituted by hidden structures, including subjective, representational, and imaginative structures.

Lastly, I want to support naïve realism. Thus an account of perceptual experience that allows perceptually apparent structure to be actual structure is preferable to a theory that does not.

Since I use these assertions in my explanation of perceptual experience, I assume I can use these claims in the explanation of candidate hallucinations in terms of something else. With that in mind, I suggest this: Similar to how we can explain perceptions of external things even with a time lag, we can re-interpret much of the evidence for candidate hallucinations as evidence of a time lag. The resulting erroneous experience is either illusion or anosognosic experience.

10.4.1. Eliminative Perceptual Theories of Hallucinations

I suggest two alternatives to an imaginary or indirect memory re-description of hallucinatory experiences. These are what I shall call *direct realist* theories. There are already direct realist and direct perception theories in the literature. The theories are comparable to those theories (perhaps even identical). In any case, why I refer to these theories here as direct and realist is because they have the following features: for some x, these theories concern an attitude (e.g., a perception, thought, or experience) towards x and hold the following about that attitude:

(a) The attitude towards x is not through something else (e.g., an internal object or property, as in sense-data theory, or through a separate thought, as in inferring one thing from another). The experience is *directly* of x.

(b) The attitude towards x is not merely of an intentional object (e.g., an imagined object, a represented thing, or a thought about non-existent creatures). The experience is of a real x.

As such, a direct realist theory has the following relationship to naïve realism: Naïve realists hold that an object of an experience partially constitutes the experience, and there is a perceptual relation between the subject and the object. Direct realist theories allow both of these: The candidate hallucinatory object is real and can partially constitute the experience; the perceptual relation is the direct relation between an object and subject.

As such, given evidence of candidate hallucinations, the following distinguishes these direct realist theories to the alternatives:

(1) Other theories hold that that there is nothing there—or if there is, we cannot experience it. As such, there only seems to be something there.
(2) Direct realist theories hold *that there is* something there and we veridically experience *that something there*. As such, there is something there and there seems to be something there.

Again, there can still be an erroneous experience here. It just is not an erroneous experience of the object itself. We can still fail to experience something else related to it, such as its properties or relations.

The direct realist theories considered here are the *direct memory theory* and the *direct perception theory*. Both hold that candidate hallucinations are direct veridical experiences of something real. The erroneous experience is something else.

The memory theory holds that there is a direct memory experience of objects, and this is what explains the evidence for candidate hallucinations. For example, when a PTSD sufferer reports a flashback of a traumatic event, the memory theorist holds that they do experience that event, and do so without mediation of something else. In naïve realist terms, a remembered object partially constitutes the memory experience.

The significant difference between this theory and the previous memory account is that this theory is a commitment to memory experience being an unmediated experience. This may correspond to the phenomenology but runs counter to the view of memory as itself being indirect, a view motivated by distinguishing memory experience and perceptual experience. If memory were indirect, this would not be a direct realist theory: The object is actually a constituent of a previous perceptual experience; 'through memory' means that the experience is indirectly related to its object.

The way to avoid this is to develop a direct theory of memory, one that is of past things without mediation of a previous perception. There are direct theories of memory. Debus (2008) defends such a theory, arguing that past objects are constituents of the memory experiences of them. Locke (1971) investigates the possibility of such a theory, but concludes it is not successful.

Yet, even if a direct memory theory is successful, it still seems to have the following problems of the previous memory account. The experience in candidate hallucinations does not have the same phenomenology as a memory experience; it has the phenomenology of a perceptual experience. Unless one dismisses the phenomenological differences between different kinds of experience (a dismissal over which anyone sympathetic to naïve realism ought to hesitate), this requires some account of how memory experiences can have the same phenomenology as perceptual experiences.

This is not a fatal flaw of the theory. Perhaps we can explain candidate hallucinations in terms of direct memory in a way that we cannot in terms of indirect memory. To do this, I must address a number of issues. The alternative to a memory theory, a perceptual theory, shares these issues. As such, given the focus of this book, I do not develop a theory involving direct memory any further. Instead, let us move on to explaining candidate hallucinations using perceptual theory.

10.4.2. Delayed Perceptual Experience

How can a naïve realist theory of perception explain hallucinations? It does so by exploiting some features already assumed in the structure of perceptual experience under tenseless eternalism. Those features of perceptual experience are

(1) Causal relations between its elements (relations that can be intrinsic to the experience);
(2) Temporal and spatial structure (including location and extent);
(3) Complex spatial relations between its subjective activity and other elements, including things located at different places and at earlier times (that is, times which are past to its subjective activity); and
(4) Illusory and anosognosic experiences of many properties and relations between perceived things.

With those features in place, I propose the following account of evidence that allegedly supports actual hallucination. This is an account that does not require actual hallucination.

First, this is a direct perceptual experience of external objects. The apparent (external) objects partially constitute the experience, without mediation of anything else.

However, this perceptual experience is different to common perceptual experiences in an important way. The *subjective activity* of this perceptual experience takes significantly longer than the more usual instances of perceptual experiences. By 'significantly', I mean that the subjective activity takes so long that, by the time it seems to the subject that they are experience it—e.g., the time when they are interacting with others—no one having a more usual perceptual experience could be stimulated by these objects.

I call subjective activity with such a temporal difference in it a *laggy* or *delayed* subjective activity or a *subjective lag*. I call a perceptual experience with such a temporal difference in it a *subjectively laggy* or *delayed* perceptual experience. This subjective lag is what gives rise to the interpretation of the experience as being of something that *is not there*—i.e., as a hallucination:

- Subjects having the candidate hallucinatory experience perceive what is *there* at a time earlier than it appears to be due to a combination of both an objective time lag (e.g., the time light takes to get from distant stars to their sense organs) and the subjective lag.
- Other subjects who do not have the candidate hallucinatory experience perceive what is *there* at a time earlier than it appears to be due only to the objective time lag.

As such, the subject perceptually experiences an object being there that the others do not experience as being there.

Note that the candidate hallucinated object is not significantly past to the subjective activity taken as a whole. The lag is in the subjective activity. What is 'significantly past' to it depends on what stage of the activity is selected.

For example, if the candidate hallucination is of my grandmother as she was 30 years ago (to use an example from Manzotti 2008), in this account, it is possibly because it takes the subjective activity 30 years to complete. This subjective activity is much longer than any normal perception of the external world. As depicted in Figure 10.1, we have the following situation.

Relative to the beginning of the subjective activity, the object I perceptually experience can be milliseconds ago—e.g., the time (t_1 to t_2) of the objective activity, in this case of the light reflecting from my grandmother's face (HXS_x at t_1) and traveling to strike my retina (A at t_2).

Relative to the end of the subjective activity, the object I perceptually experience is decades ago—e.g., the duration (t_2 to t_8) of the subjective activity that leads to the experience of my grandmother (E at t_8), and whatever is happening when that happens—e.g., I report my experience to worried friends (B at t_9).

What matters is that, as with common time-lagged perceptual experiences, it seems to us as if what we perceptually experience seems to occur at the same time as everything else, including our subjective activity. The apparent simultaneity of all of these events with the object of the delayed

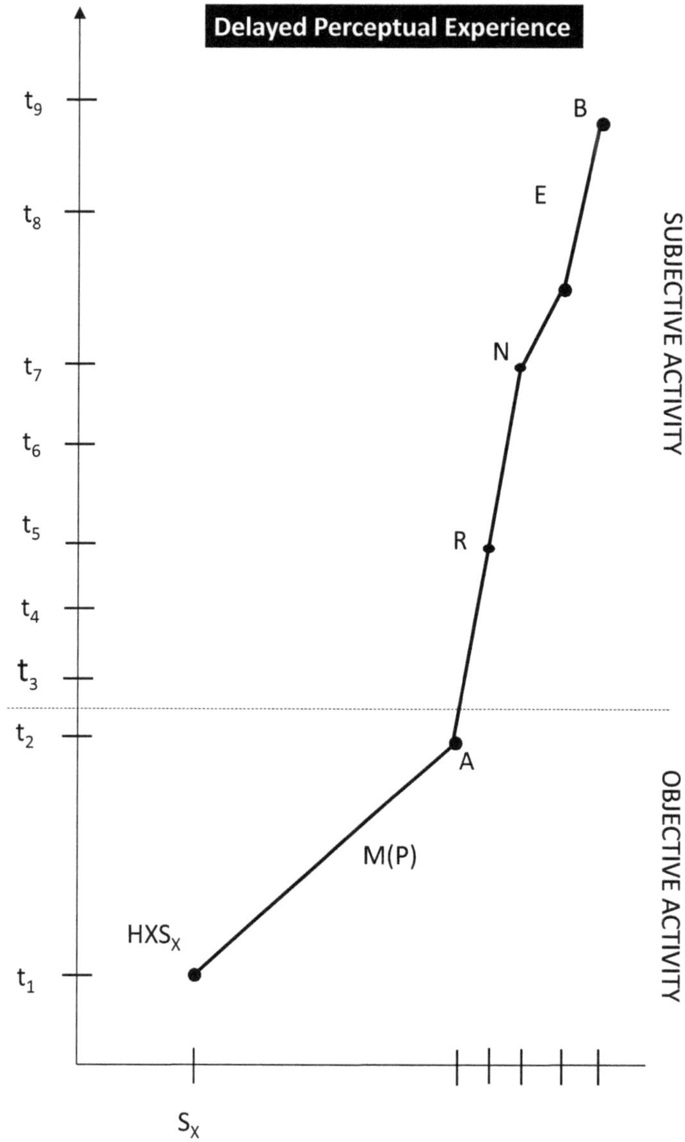

Figure 10.1 Delayed perceptual experience, with objective lag and significant subjective lag

perception is why it seems to be an object that, given our other ways of checking (e.g., trying to touch it, asking others), is not there.

However, there is still error here. Similar to the normal time-lag case, it seems to the subject what they perceive happens simultaneously with everything else they experience or think. This is mistaken. The perceived thing is at a much earlier time. However, as with the normal time-lag case, this is not a hallucination. There are objects and we do perceive the objects. The mistake is that we are anosognosic of the duration between the objects and subjective activity, and within subjective activity itself.

This brings up the next stage in the elimination of hallucinations. Our perceptual experience is mistaken because there is a *conflationary error* in our perceptual experience of multiple objects.

10.4.3. Conflationary Errors of Multiple Past Objects

Objects at multiple times (or a certain number of times) can seem to be at one time (or a lesser numbers of times). Again, distant stars can appear to shine at the same time, and that time can appear to be the same time as the subjective activity of my perception of them. Yet due to time lag and distance, they shine at different times to each other and to my subjective activity.

Furthermore, we can have perceptual experiences of only some parts or properties of objects, and have those combine.

This can explain a range of unusual perceptual experiences, including of things that do not or have never existed. Objects at these multiple temporal locations can be conflated together, appearing to be at the same time. If a subject perceptually experiences these apparently simultaneous objects as being at the same location, or as significantly spatially related, these objects may appear to be the same object or to be parts or components of the same object.

This is not a hallucination, but it is a conflationary error of what we perceptually experience: What we perceptually experience erroneously appears to be a combination of objects, properties, or relations that it is not.

Thus, we can have conflationary errors, anosognosia, and sortal illusions. However, none of these is hallucination.

To take a fantastical example, say that I have a perceptual experience of a winged human (and so, incidentally, I may interpret them as being an angel). I am in company, and no one else perceptually experiences it. I may conclude that I am hallucinating.

An alternative explanation is as follows:

(1) I am having a delayed perceptual experience of a person and of wings.
(2) The perceptually experienced person is at one time and the perceptually experienced wings are at another time.
(3) Relative to me, the perceptually experienced person is at the same place as the perceptually experienced wings. Both person and wings are *there*.
(4) The person and wings perceptually appear to be *there*.

(5) The person and wings do not perceptually appear to be at different times.
(6) The perceptual experience of the wings might be a perceptual experience of a bird, but not of all of the bird, but only its wings.

As such, it appears to me that there is a winged person *there*.

We can do this for any perceptual experience of objects that we could never have encountered. So long as the appearance of candidate hallucinatory things are decomposable into separate appearances of their wholes, parts, properties, and relations, then we can have a delayed perceptual experience of them.

Let us go back to an example at the beginning of the book. My neighbour reported seeing Saint Anne and knew it was she because she looked like her portrait. We can analyse that as a delayed perceptual experience of a picture of Saint Anne, but one which is not of the whole picture, and so seems to be of a person who 'looks like' (but is not a portrait of) a portrait of Saint Anne.

10.4.4. A Proposal to Test for Delayed Perceptual Experience and Subjective Lag

It is important to stress that this proposed analysis is a hypothesis. This is a speculative chapter. I do not believe that it is true, only that it *might* be true. However, I suggest a general way we may discover if it is true.

For some reason (empirical, conceptual, metaphysical), we may settle on what the subjective activity can be for any possible perceptual experience. In doing so, we may discover that such subjective activity cannot have so extreme a lag as required for my proposal here. If that is the case, then a re-interpretation of evidence allegedly of hallucination in terms of subject lag is not available. When I perceptually experience an object that I cannot perceive (even compensating for an objective time lag), then we are back to a case of hallucination.

If one does deny the possibility of a lag, then they need to provide some argument or evidence for this denial. It does not seem impossible that the subjective activity of a perceptual experience must be invariant in its temporal extent (or spatial extent, for that matter). It is not obvious that a perceptual experience is limited to any particular scale. If it can vary this way, then we might have reason to think that what seem initially to be hallucinations are, in fact, cases of perceptual experience with a significant subjective lag.

If there can be significant subjective lag in perceptual experience, then I think there is a way we can have evidence of that lag.

This requires a specific kind of account of what happens physically during the subjective activity. This account is of the structure of the activity as it is independent of scale. Thus it is something like what is depicted in the figures in Chapter 6 or earlier in this chapter. Although differences in temporal points (and spatial points) are marked in the figures, they are not proportional or to any predefined scale. For example, the difference between t_1 and t_3 in the figures could be any period of time. These figures only depict aspects of physics or experience that are not a matter of magnitude or scale.

If we examine only this aspect of subjective activity in what we initially take to be hallucinating subjects, then we may discover an abnormally lengthy instance of it. For example, when we examine subjects suffering candidate hallucinations, we find the same aforementioned structure as in non-candidates. The difference is that, in the former subjects, there is a much longer period between the subjective activity's beginning and completion.

This leads to an observable difference between these kinds of experiences. We use the fact that they are different in terms of scale. If there is delayed perceptual experience in cases some theorists would call hallucinations, then we observe the subjective activity during a period of time in which we the whole of the subjective activity usually occurs. If there is a delayed perceptual experience, then we only observe a stage of the subjective activity.

Figure 10.2[3] depicts the difference in subjective lag between delayed and normal perceptual experience.

During the period between t_8 and t_9 (the highlighted area in the figure), we observe subjective activity in two instances of perceptual experience: one is laggy activity in delayed perceptual experience and the other is normal activity in normal perceptual experience (called just 'perceptual experience' in the figure). With scale and proportions put aside, both experiences have identical subjective activity (A, R, N, E, and B). However, during the time we observe them, the evidence of their subjective activity is very different. There is a great deal less happening during that period in delayed perceptual experience than in normal perceptual experience: In delayed perceptual experience, only stages E and B happen; earlier stages are past. In normal perceptual experience, stages A, R, N, E, and B happen.

Finally, if there is no empirical evidence against it, then, given eternalism, I can see no reason to deny that laggy variants of subjective activity (activity with just very long durations) might lead to a perceptual experience that is at a notably different time to the stimulus that gives rise to it. In that case, it does seem as if the perceptual experience would appear to be a hallucination. For everyone around the subject and to the subject's own investigations of their environment (e.g., reaching out and touching things), it is an appearance of something that is not there.

Still, for some candidate hallucinations, it may be there is no possibility of delayed perceptual experiences being constituted by a sufficient range of perceptual objects, properties, and relations such as to capture

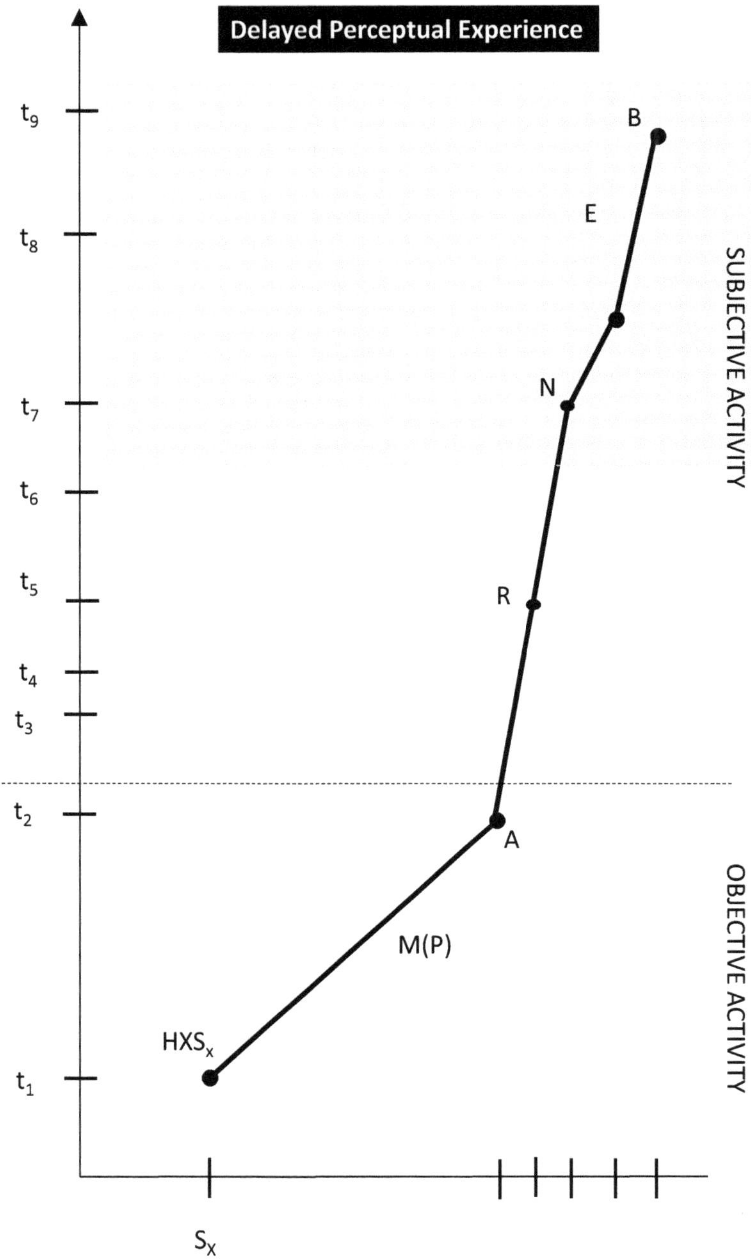

Figure 10.2 Comparison of subjective lag in delayed and normal perceptual experience: Delayed Perceptual Experience

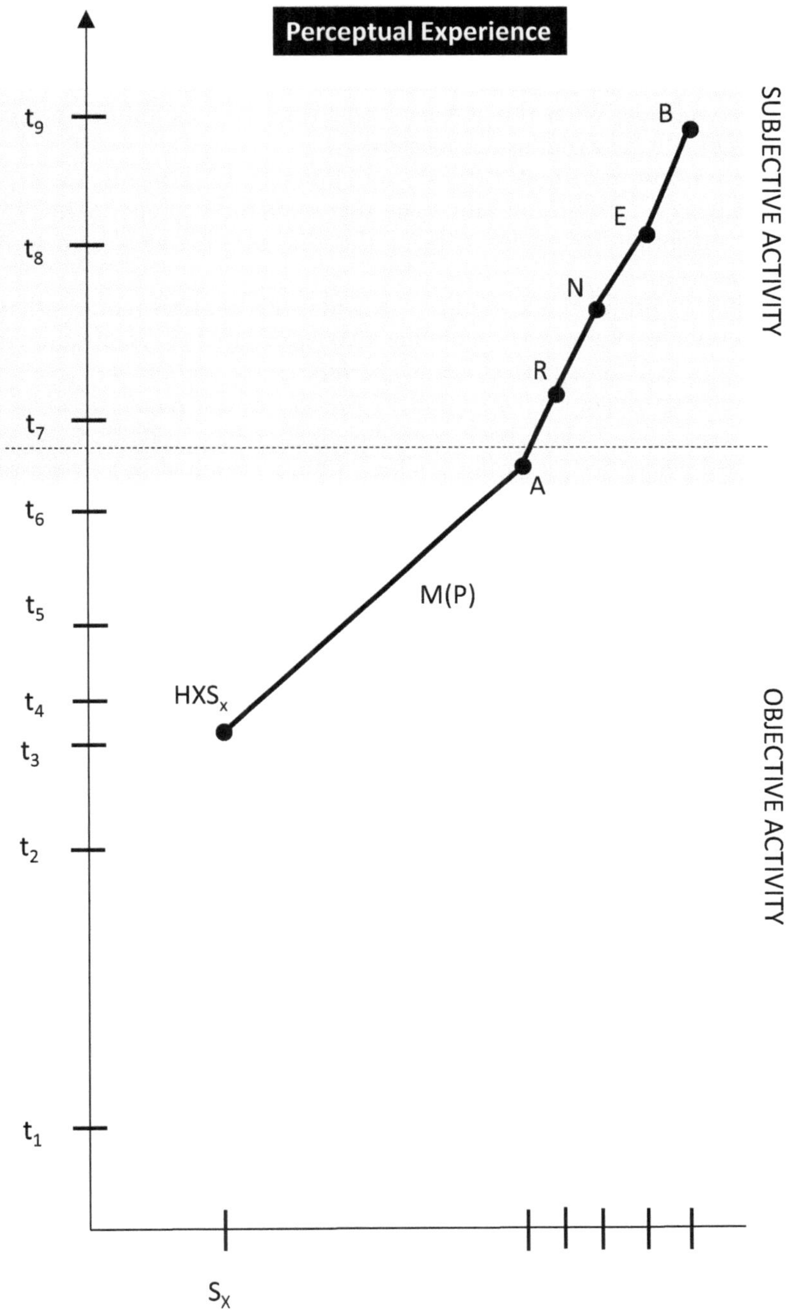

Figure 10.3 Comparison of subjective lag in delayed and normal perceptual experience: Perceptual Experience

the phenomenology of candidate hallucinations. For example, take after-images. They arguably appear to be of some thing that is no there. Can we understand them as the result of delayed perceptual experience? This seems unlikely. What things have I seen that, even as a dimly perceived collection, can have those properties?

Again, I do not think that we can eliminate all erroneous perceptual experiences. Yet this does not mean that we should accept an analysis of any perceptual experience as being the kind of error attributed to it on first reflection.

Another problem one might raise is that this analysis requires that we perceptually experience something that stimulated us at another time. One would assume that, closer to the time of the stimuli, we have a perceptual experience of the stimuli. However, we can have candidate hallucinated things which have never been experienced before. If we posit delayed perception, then the original stimulus resulted in no perception at that time. Why did it not?

My answer is that I do not know why delayed perceptual experiences happen. My intent here is only to sketch out how they can happen.

In any case, a subject might very well have normal (non-delayed) perceptual experiences at that time as well as the later delayed experience. At this stage of conception, I can see no reason to deny that both of these can occur for one stimulus. If they do, there are a number of possible experiences when the subject has the delayed perception:

(1) The subject has a delayed perceptual experience without awareness of having had an earlier perception. In this case, it seems to the subject that they have never had a similar experience.
(2) The subject has a delayed perceptual experience and is aware of having had an earlier perception. In this case, it seems to the subject that they have had a similar experience.

Option (2) may be what happens to sufferers of PTSD. The subject has a delayed perceptual experience of a traumatic event while also remembering that they once perceived it in the past. One would have a similar experience if they experience a loved one in a familiar situation after that loved one's demise.

10.5. Examples of Eliminating Hallucination with Subjective Lag

To close, let us look at how delayed perceptual experience (or just delayed perception) might deal with some examples of candidate hallucinations. We might hold that there are hallucinations in perceptual experiences of multiple things that are not there, dreamt things, and unreal things. All are candidate hallucinations because a standard way of interpreting them is that they are experiences of objects that appear to be there but are not there.

(a) *Multiple things that are not there*: Sacks gives examples of hallucinations that includes subjects with visual experiences of strange people and multiple instances of the same person in different locations (Sacks 2012). We can explain this without appeal to hallucination.

Concerning the strange people, we can explain it in one of two ways. First, these are perceptual experiences of people at other times. We have delayed perceptual experiences of them. Second, these are either people who we do not know or people we do not recognise. In both cases, we are not hallucinating them but having different kinds of anosognosic experiences of them. We have an anosognosic experience of the lag (as standard in this analysis) and then, in the case of failing to recognise a familiar person, we are anosognosic of their identity.

The more interesting challenge in Sacks's examples is that of 'hallucinations' that are of people in multiple locations. This is also explicable given delayed perceptual experience. The subject can erroneously experience multiple people at multiple places at the same time. In this case, they veridically experience people at multiple places. However, they fail to experience them as being at multiple times. Similarly, they might even veridically experience the same person at multiple places, but fail to experience them at multiple times. Thus one person may seem to occupy multiple locations.

We can expand this account to a perceptual experience of any multiple of objects, not just people. By having delayed perceptual experiences of objects and their properties at different times and places, yet being anosognosic in various ways of those different times and places, we can have any combination of these objects and properties in space. For example,

- Multiple objects can appear to be in the same place. They are in the same place, but at different times. We are anosognosic of the temporal difference between them.
- One object can appear to be in multiple places. It is in multiple places, but at different times. We are anosognosic of the temporal difference between that single object's locations at those times.

We can also run similar analyses with experiences of properties or relations. The main point is that, in all these cases, we do perceptually experience objects, but our perceptual experience fails in other ways.

As an additional point, as with temporal location, we can also fail to detect spatial differences between things—e.g., two objects that are at different spatial locations may appear to be at the same spatial location. Again, this is a form of anosognosia. We are not under illusions of spatial properties or hallucinations of objects. It is that an actual spatial difference is not apparent to us, and so we are anosognosic of it.

My point is that, in explaining our perceptual experience, given eternalism, we are entitled to appeal to a perceptual experience that has structure that is temporally extended and spatially extended, but without all that extension being apparent to us. We can use this to deal with other candidate hallucinations as well.

(b) *Dreamt things*: Assume that, when we dream, we have experiences and that those experiences seem to be perceptual experiences. How can we explain those without some appeal to hallucination?

Many dream researchers argue that our dreams seem to be constituted by objects and properties encountered in one's waking life; dreaming is a 'consolidation' of memories (Payne & Nadel 2004). Although the entirety of the contents of a dream may not be the same as anything we have encountered, the various elements, properties, and relations of the dreams are things we have previously perceptually experienced.

A theory of delayed perceptual experience explains such dreams just as it explains experiences of multiple objects, properties, and relation. It is a perceptual experience of multiple earlier objects and events, but ones in which we fail to detect differences in temporal or spatial location.

(c) *Unreal things*: There are candidate hallucinations of angels, ghosts, purple spots on the ceiling, and other unreal things. Just as with dreams and multiple figures, this theory holds that these are delayed perceptual experiences of objects, properties, and relations at different times and places, yet we are anosognosic of their temporal and spatial locations.

There is a way to challenge these suggested analyses of candidate hallucinations: one argues that nothing in the world at all resembles what is perceptually experienced. What is experienced is nothing like anything real in terms of its either elements, properties, or relations. As such, there cannot be delayed anosognosic perceptual experiences of a combination of its elements, properties, or relations.

In that case, one does have evidence of hallucination (or, perhaps, an illusion of a collection of properties). However, given the context of this discussion, one must show a reason to hold that there is or there can be an actual perceptual experience like this. It is not enough to posit a merely possible experience.

10.6. Conclusion

As stated at the beginning, this chapter presents a speculative theory, one that may seem counter-intuitive. However, as can be seen, it does not appeal to much beyond what is found in the analysis of time lag and eternalism.

The main difference is the location of the lag. In common perceptual experiences, the lag is between the distal stimuli and the subjective activity. In candidate hallucinations, the lag is between elements—i.e., in the structure—of the subjective activity itself.

One may disagree with there being a lag in the structure of subjective activity. However, if subjective activity is physical, then there is likely to be some lag in this structure. If there is no lag, then subjective activity has static structure, with simultaneous physical elements. As discussed, given those elements are spatially separated, such simultaneity, being arbitrary, is only accidentally relating elements of the subjective activity. As such, it is no constraint on the structure of the subjective activity, and so is no block to a subjective lag.

The only other difference is magnitude: The lag here can be significantly larger than that of a typical and veridical perceptual experience. It can be minutes, days, weeks, even years. One might object that no perceptual experience can have such a temporal extent in its subjective activity. Certainly, perceptual experience can have subjective activity of a few seconds or even minutes, but longer is too much.

I cannot satisfactorily defend my account from this final objection. It involves thinking about constraints based on scale or magnitude, and I have not developed any general theory on that matter. In my opinion, constraining the structure of something so varied as perceptual experience to a particular scale or magnitude strikes me as requiring some special justification. It does not seem to me that experience must have any particular spatial or temporal properties. However, this is only my opinion (most likely based on an intuition) so I leave the discussion here. In doing so, I leave two questions for those who oppose my account because of worries about magnitude:

(1) What is the upper temporal boundary of the duration of subjective activity?
(2) Given the answer to the first question, why is that duration the upper temporal boundary of subjective activity?

Given eternalism, there is no temporal limit to the reality of physical structure. Therefore, eternalism alone gives no reason to think there is a temporal boundary (presentism does have an upper limit; it is any non-zero value at all). As such, assuming eternalism, one needs some further reason to limit subjective activity to a duration. I leave others to provide the reason.

Notes

1. For a compendium on psychological concepts related to imagery, see Roeckelein 2004.

2. In Chapter 3, I argue that accounts of amodal completion as imagination have problems; obvious imagination requires an apparent unreality, whereas amodality requires a reality that is apparent yet hidden (whatever that means).

   Yet imagination works for amodality at least insofar as experiences of hidden or absent things are not like the perceptual experiences of them. Subjects are aware that perceptions do not merely present them with the amodal aspect of things. To hold that the apparently hidden aspects of an object constitute an experience is less compelling than that the perceptually apparent aspects of the object do (as would be the case if we invoke imagination to explain hallucinations).
3. As with Figure 10.1., Figure 10.2 is adapted from Figure 6.1.

11 Conclusion

This ends my examination of the structure and errors of perceptual experience in light of different philosophical positions about time. There are a number of consequences from the arguments presented in this examination. There are also a number of possible objections against assertions common to all these chapters. This conclusion sketches these consequences and objections.

11.1. Summary of Consequences

The general consequence of these arguments is that different philosophical positions on time lead to different ways of thinking about the structure of perceptual experience. Different philosophical positions on time offer different kinds of accounts for the following:

(a) *Naïve realism, by means of its particular commitment to the relationship between perceptual content and perceptual constitution.* Presentism about perception raises a number of problems for naïve realism, in particular regarding spatial relationships between the subject and perceptually appearances. Eternalism, especially tenseless theory, does not raise these problems.
(b) *Different models and structures of actual perceptual experience.* Presentism about perception requires that any real structure of perceptual experience (rather than, say, apparent or intentional structure) must be present and at one time. Eternalism does not require such a constraint on the real structure of perceptual experience; it can extend over multiple times (although tenseless theory complicates this account). This has an impact on models of perceptual experience, as evidenced in this book by models of spatial experience (including subjective activity and objectivity) and of temporal experience.
(c) Different interpretations of actual perceptual experience as veridical and erroneous. The relationship between appearances and reality is different under different philosophical positions of time. Presentism about perception makes a great deal of perceptual experience veridical,

including the perception of spatial relationships between external things. Eternalism does not.

(d) Where perceptual experience must be erroneous, how it is erroneous. Presentism makes a great deal of erroneous perceptual experience illusory or hallucinatory. Eternalism does not, but instead makes it anosognosic.

Furthermore, I make a distinction between different kinds of erroneous perceptual experience: hallucination, illusion, and anosognosia. The difference between kinds of error included a difference in their desirability. I argue that some are preferable to others: anosognosia is preferable to illusion and hallucination, and illusion preferable to hallucination.

In addition, I argue that there are other differences important to understanding the relationship between time and perceptual experience. These are differences between: (a) content and constitution, (b) embedded and extended, (c) subjective activity and objective activity, and (d) relative and absolute.

Finally, I argue that the epistemological value of actual perceptual experience (its trustworthiness) can vary depending on positions of time. How things perceptually appear, the phenomenology of perception—whether or not this corresponds to how things are, to the ontology, depends on whether presentism, eternalist tense theory, or tenseless theory is true.

I develop, illustrate, and defend this position by analysing the perceptual experiences of space and time and, in particular, considering them in light of time lag and illusory depth.

The main conclusion is that if one prefers having trustworthy perceptual experience, then one should prefer eternalism to presentism and tenseless theory to tense theory.

As such, the arguments of this book have consequences for choosing theories in two relevant philosophical areas of research: the philosophy of time and the philosophy of perception.

11.1.1. *Philosophy of Time*

If these arguments work, then perceptual experience implies tenseless theory and eternalism are preferable to their alternatives of tense theory and presentism. In this book, I argue on a case-by-case basis for this position.

A more general question is this: What is it about tenseless theory and eternalism that make them better than their alternatives? It is not their alleged counter-intuitiveness; it is not their complexity, unfamiliarity, or relationship to physics. Here I offer only a speculation; there is scope for more work on this.

In comparison to competing positions, many physical theories hold there to be less in the world. For example, in the philosophy of mind, physical *monism* holds that there are only physical things, whereas non-physical *substance dualism* holds there to be more—physical things (like monism) but also non-physical mental things. In the philosophy of life, *biology* holds that living things are only complex physical things, whereas *vitalism* holds that living things are physical things which also contain something particular to life—'animal spirit' or *elán vital.*

The relationship between the physical theory of time and its competitors is not like this. In the philosophy of time, the physics of time both reduces and expands the number of real things in comparison to competing positions.

In comparison to the consequence of physical theory that is tenseless eternalism, presentism holds that there are fewer things than eternalism. If eternalism is consistent with physical theory, and presentism is not, the more expansionist theory is consistent with physical theory. However, this is not a one-dimensional difference: Tenseless theory also reduces the number of absolute temporal properties. It only accepts relative tenses, and these are derived by temporal relations (e.g., earlier, later, simultaneous with). In contrast, presentism and other tense theories hold that there is a special present (and past and future), not merely tenses derived from tenseless temporal relationships. Furthermore, tense theory, unlike tenseless theory, holds that there is temporal passage.

I have argued that a tenseless theory ontology captures the phenomenology of perceptual experience better than a tensed or presentist ontology because it does two things:

(1) Where it expands the ontology in comparison to its competitors, it captures more of what is apparent to us than tense theory or presentism—e.g., distant shining stars that are present and real. Tenseless theory allows both for the apparent stars, despite their being at a different time to local present (and real) subjective activity.
(2) Where it reduces the ontology in comparison to its competitors, it often does so by relativizing the relevant properties—e.g., presentness, pastness, futurity. Where this relativity conflicts with the phenomenology of perception, we can explain the difference by anosognosia. We experience a tense, but not the tense's relativity (or indexicality); by not appearing to be relative, the tense appears to be absolute. Because it needs something to be relative to, there is more to a relative property than there is to an absolute property. As such, where a relative property appears to be an absolute property, there is more in the world than there appears to be.

The only exception to this, I think, is temporal passage. As such, the phenomenology of temporal passage remains a challenge for tenseless theorists. As a tenseless theorist, I attempt to meet this challenge in the chapter on temporal experience; however, I do not assume that my proposed solution is sufficient for the doubtful or undecided.

11.1.2. Philosophy of Perception

I avoid directly addressing the variety of competing perceptual theories. I do not weigh them against each other; certainly, I do not weight them independently of the philosophy of time. I focus on naïve realism, in part because it is a position that ought to be familiar to any perceptual theorist.

Turning to naïve realism, one of the motivations here is the epistemological role of perceptual experience, and, in particular, its role in our knowledge of the actual world. We can know about the actual world by experiencing it and by that experience being a case in which how things appear is how things are.

The arguments of this book, then, have been that the viability of naïve realism depends on how we think about time. As such, although perhaps indirect, these arguments can contribute to the perceptual debate. If the arguments of this book are correct, then, given tenseless eternalism, naïve realism is in a better position than it is if other positions on time are true. At least, naïve realism is in a better position for specific kinds of perceptual content. As discussed in the chapter on depth, as far as I can tell, naïve realism must be false for the perceptual experience of depth.

Finally, it is worth noting that physics motivates tenseless eternalism. As such, this is an occasion in which a physically motivated and allegedly counter-intuitive theory supports the appearances, but its competitors do not.

Now consider some general objections to the project.

11.2. General Objections

I can think of four general objections that we can raise against this project:

(a) Perceptual experience is not important for knowledge of the actual world.
(b) The epistemological worries about illusions and hallucinations are exaggerated.
(c) The philosophy of time makes no difference to these epistemological worries.
(d) Finally, one particular to the tenseless theory view of perceptual constitution developed here: causal relations cannot be intrinsic to a perceptual experience and/or cannot hold between its constituents.

11.2.1. Perceptual Experience Is Not Important for Knowledge of the Actual World

The worry about knowledge of the actual world is unfounded. We can have knowledge of the actual world without perceptual experience being trustworthy. We have some non-perceptual source to justify beliefs about the actual world in such a way that we can call these beliefs knowledge. Perhaps there is some kind of 'Moorean' belief (to use Crisp's (2005) term); perhaps there is something everyone believes, something there is no reason to doubt. This belief is about the actual world. As such, we can have knowledge of the actual world without relying on trustworthy experience.

For example, we believe that the actual world contains external things. We may argue: This is an indefeasible 'Moorean' belief sufficiently formed or grounded to be always a kind of knowledge.

This may work for some beliefs about the actual world. However, I do not think it works for all of them. There are many beliefs that seem to lack foundation in the absence of some instances of experience (be they those of the believer, another individual subject, or of multiple subjects' experiences). For example, there are beliefs about the particular ways subjects move through the external world shape, constitute, or relate to it. If it is possible to gain such knowledge without experience, I do not know what such knowledge could be.

Indeed, one might even apply this to belief that there is an external world. Why do I believe that this is a kind of fundamental knowledge? Why is it not as plausible that I am a disembodied spirit or an occupant of some kind of non-spatial realm?

My own answer from intuition is that I have experiences that show me parts of this actual world, parts that are external to me and located in external space. If I did not have such experiences, or could not trust them, I do not know what belief would be a suitably justifiable alternative.

Given the prevalence of empiricism as a response to rationalism with respect to the actual world, I am not sure anyone else has come up with an alternative kind of belief or source of beliefs. To find one would be an interesting project to pursue, but until it is pursued and found, its mere possibility is no objection to holding that experience is an important source of our knowledge of the actual world.

Furthermore, even if there is such an alternative source, what guarantee is there that the same problems do not arise in this alternative? If they do arise, then there are similar problems about this alternative source of knowledge. All we do is substitute one source of doubt for another.

11.2.2. The Worries about Illusion or Hallucination Are Exaggerated

One might hold that, even if perceptual experience is erroneous by being illusory or hallucinatory, this should not be troubling. Some perceptual experiences are erroneous; others are not erroneous. This is well known, an old and significant part of epistemology. Yet we still have beliefs about the actual world; we still claim knowledge of it. Something must allow those two states of affairs to co-exist.

One might think this because one assumes that all actual perceptual experience can have instances that are illusory or hallucinatory, and yet one believes that beliefs about the actual world are justified. There is no conflict here.

Why is there no conflict here? If my experience can occur in the absence of what seems to constitute it, then what is apparent through it is not sufficient evidence of that apparent thing. If I can perceptually experience trees without there being trees, then my perceptual experience of trees is not sufficient evidence for there being trees.

One way around it is to assume that, every time there is an appearance of trees, there are trees. This does not work for everything. However, it can work for some experiences—and so some experiences can be justified.

An assumption of this project is that not all veridical experiences need have illusory or hallucinatory counterparts. This must be shown either on a case-by-case basis (e.g., we show there are particular illusions of colour), or shown for some fundamental properties, such that actual illusions of these properties means there can be actual illusions of all properties.

Nevertheless, perhaps I do exaggerate worries from erroneous experience. The point of this project is that, for some contents of perceptual experiences, given certain ways of thinking about time, these worries need not arise.

11.2.3. *The Philosophy of Time Makes No Difference to These Epistemological Worries*

The philosophy of time does not make a difference to the epistemological worry. We still have error—we just have a different kind of error. This commitment to other error is just as bad for the trustworthiness of perceptual experience in forming our beliefs about the actual world. As such, the arguments here do not motivate a preference for one theory over another.

For example, I see what appears to be a distant real and shining star. If presentism is true, this apparent star shine is unreal. If eternalism is true, the apparent event is real but at a different time. Both of these are equally bad. Perceptual experience is still mistaken. By basing our beliefs about the actual world on this experience, they remain unjustified beliefs. Given the role perceptual experience plays in how we form beliefs about the actual world, it hardly matters how they vary in this error. Once combined with other sources of beliefs about the actual world, the contribution from the difference between the kinds or error is barely relevant.

I hope I show that, on a case-by-case basis, thinking from different positions about the philosophy of time affects commitment to different kinds of error in perceptual experience. For beliefs about the actual world, if some of those kinds of errors are worse than other kinds of errors, then these different positions are worse for such positions.

For example, if we can have hallucinations of external objects (or illusory externality), then we do not have a trustworthy appearance of external objects and so evidence of an external world. If we can only be anosognosic of external objects, then any appearance of an external object is a trustworthy appearance of the actual external world. Anosognosia requires that there be something external out there that we mistakenly experience in some way. (Of course, we may doubt there is anything external at all; however, this has no place in any realist models of perceptual experience.)

Another objection is that, perhaps, this focus on perceptual experience does not matter very much; thus, philosophy of time's effect on perceptual experience does not matter very much. Perceptual experience contributes to our beliefs about the actual world. However, any particular perceptual experience we are having does not matter. This is because it is combined with other kinds of experience and cognition, and the errors in those other kinds undermines beliefs about the actual world far more than perceptual experience.

For example, I believe in part that there is an actual external world because I perceptually experience external things. However, I predominantly believe that there is an external world because I perceptually *experienced* external things; I had many past perceptual experiences. How do I know I perceptually experienced external things? I *remember* perceptually experiencing external things.

As such, my beliefs about the actual world more heavily depend on my memory of previous perceptions than the perceptual experience I am having. Even if I can trust my perceptual experience, if I cannot trust my memory, then I cannot trust a lot of what I take from experience.

If my memory is untrustworthy, then my beliefs about previous experiences may also be untrustworthy. Even if my perceptual experience can be trusted, my memory cannot. If I cannot trust my memory, then, because I cannot trust it, I cannot trust any previous perceptual experiences either.

With respect to the external world, I might reply that it does not matter. My perceptual experience that I am having, and not merely remembering, is trustworthy and is of the external world.

However, one might object further. A belief that there is an external world may satisfy the most radical sceptic. Yet, surely, I want more beliefs about the actual world than that. That belief seems quite empty and inert. What about my belief that the external world contains such things as three-dimensional shapes, slowly changing things, distant events, colours, animals, persons, people, selves, my own body, free will, ethics, valuable states such as mutual love or hope, and common states such as fear and suffering? All of these seem to turn on more than any particular perceptual experience I am having.

In response, first, this is indeed an issue. The unreliability of memory complicates matters and, unresolved, may further make perceptual experience to be epistemologically problematic. However, I think we must start somewhere. Ensuring that perceptual experience is as trustworthy as it can be seems a good place to start. If we can at least trust that we can take perceptual experience in some cases as a trusted source of belief about the world, then we can move on to engage with questions about other sources, such as memory.

Second, even if memory plays a significant epistemological role, perceptual experience in general is still more important. Memory may both supplement any single perceptual experience as another source of belief about the actual world; memory may contain more sources of belief about the actual world than any single perceptual experience. However, memory is not itself a primary source about the actual world. Its position is dependent on the perceptual experiences related to memory. Memory gets us justified beliefs about the actual world through perceptual experience.

For example, my belief that someone I meet today is the same person I met yesterday is not due only to a perceptual experience and a memory of them. It is that I remember *perceptual experiences* of them. A position on the reliability of perceptual experience affects any position on the reliability of memory.

Third, the fact memory plays a significant epistemological role means there is further work to do, work on the trustworthiness or reliability of memory—especially given work in the last few decades on memory's unreliability—e.g., false memory syndrome (e.g., Loftus 1996; Brainerd & Reyna 2005; Sabbagh 2009; Shaw 2016). For example, perhaps future research may find that different positions on time lead to different analyses of the trustworthiness of memory (similar to what I argue for perceptual experience).

Finally, another way in which one might deny the importance of time is that different positions on time still lead to error. One might argue that showing that different kinds of error occur under different kinds of experiences is irrelevant. A commitment to any kind of erroneous experience is as bad as a commitment to any other kind of erroneous experiences.

I argue why one might think otherwise—e.g., that a theory committed to anosognosia is better than one committed to illusion and one committed to illusion is better than one committed to hallucinations. However, these arguments may not be convincing.

However, note that, in some instances of experience, there is still less error under some positions on time than others are. For example, seeing a distant event, one fails to detect the duration between the distant event and one's own subjective activity. Given tenseless theory, this is all there is to the error in this experience: The event is real and as present as one's own subjective activity. One is anosognosic of the duration. In comparison, given tense theory, along with the anosognosia of duration, one is under an illusion of the presentness of that event and anosognosic of the pastness of the event. Simply due to a difference between the theories, there is more error in an experience due to tense theory than there is due to tenseless theory.

11.2.4. Causation and Direct or Naïve Perceptual Experience

A more specific objection to both the structure of perceptual experience and the philosophical position of time is this: causation cannot be a constitutive relation between elements of a perceptual experience. As such, my assertions that I can directly experience a cause of my experience and that a cause of my experience can be constitutive of my experience are false.

Yet why is it that causation cannot constitute? Is it because it is counter-intuitive or is it because it is impossible?

If it is counter-intuitive, I wonder which is preferred: the intuition that causal relata do not constitute or the intuition that much of how things appear (e.g., there are distant external events) is how things are?

We cannot have both. If causation links the external things to the subjective activity, if subjective activity partially constitutes a perceptual experience, if the external things appear to constitute a perceptual experience, and if how things seem is how things are, then causally related entities (the subjective activity and the external things) constitute the same perceptual experience. Deny causal relata constitute and you deny such appearances.

However, perhaps we must deny the appearances. It is not that one prefers that the causal not constitute the same thing, it is that it is impossible that they do.

Yet I do not know of a reason to hold it to be impossible. Furthermore, putting experience and phenomenology aside, I can think of a reason to hold that it is possible.

Perhaps no cause and effect can constitute the same thing because by 'thing', one means 'object', and a cause and its effect cannot constitute an object.

However, experiences need not be objects. There is no a priori reason why they cannot be events or processes, things that *happen*. If experiences are events or processes, then they can be temporally extended events or processes. Temporally extended events or processes can be a sequence of causally linked stages—e.g., the process of a plant growing includes causally related stages such as absorbing sunlight, creating chlorophyll, and growing.

Furthermore, even if experiences are objects, some metaphysicians do conceive of causation holding between parts or constituents of an object. This is most notable with *temporal parts*. Temporal parts can be part of an object and causally related. Their parthood explains the persistence of the object through change; their causal relationship can be the relationship between stages of one such change (e.g., Sider 2003). If you deny that a cause and an effect in a causal relation cannot constitute something, then you also deny temporal parts.

If one is a presentist or tense theorist, deny temporal parts may be an acceptable position. However, most tenseless eternalists accept temporal parts for objects (e.g., Sider, *op.cit;* at least one does not: Mellor 1998). The overall thesis of this book concerns a tenseless theory analysis of perceptual experience.

As such, given tenseless theory, I see no reason to deny that causally related things constitute a perceptual experience.

I do not consider this work the final word on the relationship between the philosophy of time and other philosophical or scientific discussions. The central question to take away from this work for any other area of research is this: Given some entity you posit as existing, how does it exist and how is it structured in time? If that is an important question, then this book has argued that how one answers it can depend on one's position on time.

If the arguments of this work are valid and their premises plausible, then they may not be restricted merely to perceptual error and naïve realism. There may be other areas of research. For example, one may wonder how thinking through different positions on time may affect other theories of perception, experience, and the mind in general.

11.3. Return to the Purple Spots

At the start of the book, I described how, as a child, I had a perceptual experience that I want to be erroneous. Do my arguments concerning the philosophy of time and the constitution and phenomenology of perception alter what I can or must believe about this perceptual experience? Must it be an illusion or hallucination whatever my theory of time? Alternatively, could it be veridical or, as preferred by naïve realists, anosognosic depending on my theory of time?

To take these experiences exactly as I describe them is a commitment to an ontology that many may find to be too much, including myself. I consider it too much to commit to the existence of gremlins.

So how can I explain my experiences in a way that is compatible with naïve realism? The way to do so is to analyse these experiences in terms of anosognosic errors.

Independent of the philosophy of time, there are a few things I need not accept about the purple spots. I need not accept my brother's testimony. Given the experience of the spots themselves, they may very well be gremlins, but that they are purple spots is insufficient proof that they are gremlins. They could be something else and the experience be the same.

One assumes that gremlins are creatures with malicious intentions. All these spots indicate is that they are in motion. There is nothing obviously malicious or even intentional about that. More proof is needed to show that these spots are gremlins (which even then I knew of course—but had no desire whatsoever for any further proof).

So what did I perceptually experience? What I experienced were purple moving spots at a distance from me. Can I then at least have that—an experience that is seeing purple spots in motion at a distance from me?

Whatever I say, there cannot have been any purple spots moving on the ceiling at the time I was having the experience, especially not any independent of me. There was nothing up there but the dark ceiling.

What one experiences need not be simultaneous with the subject and their subjective activity. So that there are no purple spots up there now does not mean that I do not experience them. If some moving purple spots are in the right way perceptually related to me—e.g., through causal chains—then I can be veridically experiencing their motion, their 'purple'-ness and their spottiness. We have the account advanced in this book: I just cannot be veridically experiencing the duration between them and myself. I am anosognosic of that duration. There appears to be no duration, but there is a duration.

However, the delay involved here cannot be the delay of light leaving the surface of something and reaching my eye. That delay is a time lag in the objective activity constituting the experience. Yet there is no objective activity corresponding to purple lights moving on the ceiling. That would require lights being there a few billionths of a second ago.

At this point, one might think there is a simple answer here: The spots, their motion, their colour are all merely a product of internal or subjective processes. I am seeing something that is dependent on me in some way; for example, it is activity within the retinal field or some related neurological activity, such as neuronal bleaching or adaptation (Zaidi et al. 2012).

Thus these are instances of subjective activity, such as sense data, or representations due to instances of subjective activity. However, this is not a naïve theory. Naïve theory requires that what we experience to be independent of the subject. Being subjective activity or the result of a subject makes it dependent on the subject. As such, if we do hold this view, then we must deny naïve realism about this experience (and take on other theories, e.g., disjunctivism (where experiences of the purple spots are 'bad cases') or representationalism).

However, there is a way one can bring in time in order to explain this experience in a way that is compatible with naïve realism. This method is a variant of the method for eliminating hallucinations.

Here is what one might say: I do indeed see something coloured and moving at a distance from me. Furthermore, drawing on the analyses of alleged hallucination, I see the coloured moving thing through a process of delayed perception. However, my perceptual experience is a very limited form of experience. It is indistinct and undetailed. It is a like a kind of 'squinting' within my perceptual faculties when looking at the world, or like having somewhat bad hearing. What is apparent in the case of the purple spots, being so indistinct, are broad and basic features such as colours and motion. What is not apparent are the many other properties the moving coloured thing possesses.

That is, the visible purple spots are due to a delayed, diminished perception in comparison to a typical perception. They are, however, something real.

Like the account in the chapter on hallucination, this kind of account is merely a suggestion. We may find it cannot work. It may be that, if we examine the phenomenological data more closely, then whatever our thinking about time, and however that data combines with other things or is related to the experiencing subjects, nothing in the real world corresponds to the experience. In that case, it seems we must have erroneous experiences such as illusions or hallucinations.

Alternatively, the kinds of solution offered here might be unacceptable in its own right. As discussed in the chapter on hallucination, an analysis of alleged hallucinations in terms of delayed perceptual experiences, may strike the reader as too much. If so, then there is nothing here preventing one from rejecting it. At least there is nothing preventing putting it aside so long as the drive to do so is due to assumptions or intuitions based on theories of time rejected here in their turn—e.g., presentism. With that possibility aside, if these solutions remain counter-intuitive, then, even if they are possible, nothing in this book forces one to accept them.

This is to be expected. The argument of this book is that different theories of time open up different ways in which experience is trustworthy. This does not guarantee saving all appearances and making all experiences trustworthy. Some are going to be erroneous, and not just anosognosic—i.e., in a way that is compatible with naïve realism. (As I argued already, the experience of depth is untrustworthy and can be illusory.)

As such, we cannot assume that experience is always trustworthy. However, unless we assume that any doubtful experience undermines all experience, or we have reason to doubt a particular experience, then at least we can assume that some experience is trustworthy.

This conclusion does not present a settled position about experience. Given the right kind of evidence, we can still show any particular experience to be untrustworthy. Any evidence that shows illusory or hallucinatory forms of that experience is sufficient.

However, if the arguments of this book are successful, then before accepting this evidence, we should check that it is not presupposing some metaphysical commitment—a commitment that, although intuitive, is plausibly false, with a better alternative to it.

Bibliography

Almäng, Jan. 2014. "Tense as a Feature of Perceptual Content." *The Journal of Philosophy* 111, no. 7: 361–337.

Alston, William. 2005. "Perception and Representation." *Phenomenological and Philosophical Research* 70, no. 2: 253–289.

Antony, Louise. 2011. "The Openness of Illusions." *Philosophical Issues* 21, no. 1: 25–44.

Arstila, Valtteri. 2003. *The Paradox of Colors*. Turku: University of Turku.

Arstila, Valtteri & Pihlainen, Kalle. 2009. "The Causal Theory of Perception Revisited." *Erkenntnis* 70, no. 3: 397–417.

St. Augustine, 1982. *Confessions*. Harmondsworth, Penguin

Bach, Michael. n.d. "88 Visual Phenomena & Optical Illusions." Accessed October 9th 2017. www.michaelbach.de/ot/index.html.

Balashov, Yuri. 2010. *Persistence and Spacetime*. Oxford: Oxford University Press.

Bardon, A. 2013. *A Brief History of the Philosophy of Time*. Oxford, Oxford University Press

Bergson, Henri. 1999. *Duration and Simultaneity*. Manchester: Clinamen Press.

Blackmore, Susan. 2003. *Consciousness: An Introduction*. Oxon: Hodder & Stoughton.

Blumson, Ben. 2009. "Images, Intentionality and Inexistence." *Philosophy and Phenomenological Research* 129, no. 3: 522–537.

Brainerd, C.J. & Reyna, V.F. 2005. *The Science of False Memory*. Oxford: Oxford University Press.

Breckon, T.P. & Fisher, R.B. 2005. "Amodal Volume Completion: 3D Visual Completion." *Computer Vision and Image Understanding* 99: 499–526.

Brentano, F.C. 1874. *Psychology from an Empirical Standpoint*. Translated by A.C. Rancurello, D.B. Terrell, and L.L. McAlister. London:Routledge, 1995.

Briscoe, R.E. 2010. "Perceiving the Present: Systematization of Illusions or Illusion of Systematization?" *Cognitive Science* 34, no. 8: 1530–1542.

Broad, C.D. 1938. *An Examination of McTaggart's Philosophy, Vol. II, Part 1*. Cambridge: Cambridge University Press

Brooks, D.H.M. 1994. *The Unity of the Mind*. Hampshire, MacMillan.

Burge, Tyler. 1979. "Individualism and the Mental." *Midwest Studies in Philosophy* 4, no. 1: 73–121.

Butterfield, Jeremy. 1984. "Seeing the Present." *Mind* 43, no. 1: 161–176.

Callender, Craig. 2008. "The Common Now." *Philosophical Issues* 18: 339–361.

Cam, Philip. 1990. "Insularity and the Persistence of Perceptual Illusion." *Analysis* 50, no. 4: 231–235.

Campbell, John. 1996. "Shape Properties, Experience of Shape and Shape Concepts." *Philosophical Issues* 7: 351–363.

Chalmers, David & Clark, Andy. 1998. "The Extended Mind." *Analysis* 58, no. 1: 7–19.

Changizi, Mark. 2010. "Response to Briscoe." *Cognitive Science* 34: 1543–1547.

Changizi, Mark, Hsieh, A., Nijhawan, R., Kanai, R., & Shimojo, S. 2008. "Perceiving the Present and a Systematization of Illusions." *Cognitive Science* 32: 459–503.

Clark, Austen. 2010. "Vicissitudes of Non-Visual Objects: Comments on MacPherson, O'Callaghan, and Batty." *Philosophical Studies* 153, no. 1: 175–181.

Clark, Stuart. 2007. *Vanities of the Eye*. Oxford: Oxford University Press.

Coates, Paul. 2000. "Deviant Causal Chains and Hallucinations: A Problem for the Anti-Causalist." *The Philosophical Quarterly* 50, no. 200: 320–331.

Crane, Tim. 2001. *Elements of Mind: An Introduction to the Philosophy of Mind*. Oxford: Oxford University Press.

Crane, Tim. 2006. "Is There a Perceptual Relation?" In *Perceptual Experience*, edited by Gendler, T.S. & Hawthorne, John, 126–143. Oxford: Oxford University Press.

Crane, Tim. 2011. "The Problem of Perception." In *The Stanford Encyclopedia of Philosophy (Spring 2011 Edition)*, edited by Zalta, Edward. http://plato.stanford.edu/archives/spr2011/entries/perception-problem.

Dainton, Barry. 2000. *Stream of Consciousness: Unity and Continuity in Conscious Experience*. London: Routledge.

Dainton, Barry. 2001. *Time and Space*. Chesham: Acumen.

Dainton, Barry. 2003. "Time in Experience: Reply to Gallagher." *Psyche* 9, no. 12. http://psyche.cs.monash.edu.au/symposia/dainton/gallagher-r.pdf.

Dainton, Barry. 2010. "Temporal Consciousness." In *The Stanford Encyclopedia of Philosophy (Fall 2010 Edition)*, edited by Zalta, Edward. http://plato.stanford.edu/archives/fall2010/entries/consciousness-temporal.

Dancy, John. 1995. "The Argument from Illusion." *The Philosophical Quarterly* 45, no. 181: 421–438.

Deasy, Dan. 2015. "The Moving Spotlight Theory." *Philosophical Studies* 172, no. 8: 2073–2089.

Debus, Dorothy. 2008. "Experiencing the Past: A Relational Account of Recollective Memory." *Dialectica* 62: 405–432.

Deng, Natalja. 2017. "Temporal Experience and the A versus B Debate." In *Routledge Handbook of Philosophy of Temporal Experience*, edited by Philips, Ian. Oxford: Routledge.

De Warren, Nicolas. 2009. *Husserl and the Promise of Time: Subjectivity in Transcendental Phenomenology*. Cambridge: Cambridge University Press.

Dennett, D. 1991. *Consciousness Explained*. Allen Lane. Harmondsworth

Droege, Paula. 2009. "Now or Never: How Consciousness Represents Time." *Consciousness and Cognition* 18: 78–90.

Engels, A.K., Fries, P., Konig, P., Brecht, M., & Singer, W. 1999. "Temporal Binding, Binocular Rivalry, and Consciousness." *Consciousness and Cognition* 8, no. 1: 206–268.

Feynman, Richard. 1971. *The Feynman Lectures on Physics Volume 1*. Philippines: Addison-Wesley.
Fish, William. 2009. *Perception, Hallucination and Illusion*. Oxford: Oxford University Press.
Furlong, E.J. 1954. "Memory and the Argument from Illusion." *Proceedings of the Aristotelian Society*, New Series 54, no. 1: 131–144.
Gallagher, Shaun. 2003. "Sync-Ing in the Stream of Experience: Time-Consciousness in Broad, Husserl, and Dainton." *Psyche* 9, no. 10.
Gibson, J.J. 1950. *The Perception of the Visual World*. Cambridge, MA: The Riverside Press.
Gordon, Ian. 2004. *Theories of Visual Perception*. East Sussex: Psychology Press.
Gregory, Richard L. ed. 1987. "Illusions." In *The Oxford Companion to the Mind*, edited by Gregory, Richard, 337–343. Oxford: Oxford University Press.
Gregory, Richard L. 1998. *Eye and Brain: The Psychology of Seeing*. Oxford: Oxford University Press.
Grush, Rick. 2005. "Brain Time and Phenomenological Time." In *Cognition and the Brain*, edited by Brook, Andrew & Akins, Kathleen, 160–207. Cambridge: Cambridge University Press.
Grush, Rick. 2006. "How to, and How Not to, Bridge Computational Cognitive Neuroscience and Husserlian Phenomenology of Time Consciousness." *Synthese*, no. 3: 417–450.
Hacker, P.M.S. 1987. *Appearances and Reality*. Oxford: Basil Blackwell.
Hawley, Katherine. 2001. *How Things Persist*. Oxford: Clarendon Press.
Heelan, P. 1983. *Space-Perception and the Philosophy of Science*. London: University of California Press.
Hilbert, D. 2004. "Hallucination, Sense-Data and Direct Realism." *Philosophical Studies* 120: 185–191.
Hoerl, Christoph. 2013. "A Succession of Feelings, in and of Itself, Is Not a Feeling of Succession." *Mind* 122, no. 486: 373–417.
Hopkins, R. 2012. "What Perky Did Not Show." *Analysis* 72, no. 1: 431–439.
Houts, R. 1980. "Some Implications of the Time-Lag Argument." Philosophical and Phenomenological Research, 157, 150–157
Humphrey, Nicholas. 2008. *Seeing Red: A Study in Consciousness*. Harvard: Harvard University Press.
Husserl, Edmund. 1991. *The Phenomenology of Internal Time-Consciousness*, trans. Brough, John. Dordrecht: Kluwer.
Husserl, Edmund. 2005. *Phantasy, Image Consciousness, and Memory (1898–1925)*, trans. & ed. Brough, John. Dordrecht: Springer.
Jackson, Frank. 1975. "On the Adverbial Theory of Perception." *Metaphilosophy* 6: 127–135.
Jackson, Frank. 1977. *Perception: A Representative Theory*. New York: Cambridge University Press.
Jacob, P. 2015. "Action-Based Accounts of Perception." In *The Oxford Handbook on the Philosophy of Perception*, edited by Mohan Matthen, 217–237. Oxford: Oxford University Press.
Jacquette, Dale. 1985. "Sensation and Intentionality." *Philosophical Studies: An International Journal for Philosophy in the Analytic Tradition* 47, no. 3: 429–440.
James, William. 1890. *The Principles of Psychology*. New York: Dover.

Johnston, Mark. 2004. "The Obscure Object of Hallucination." *Philosophical Studies* 120: 113–183.

Julesz, Bela. 1966. *Experiments in Visual Perception*, ed. Vernon, M.D. Harmondsworth: Penguin.

Kahneman, Daniel. 2012. "A Machine for Jumping to Conclusions." Interview by Lea Winerman. *Monitor on Psychology* 43, no. 2. www.apa.org/monitor/2012/02/conclusions.aspx.

Kalderon, M.E. 2011. "Color Illusion." *Nous* 45, no. 4: 1–25.

Kelly, Sean D. 2005. "The Puzzle of Temporal Experience." In *Cognition and the Brain*, edited by Brook, Andrew & Akins, Kathleen, 208–238. Cambridge: Cambridge University Press.

Kölers, P.A. 1972. *Aspects of Motion Perception*. Oxford: Pergamon.

Le Poidevin, R. 2003. *Travels in Four Dimensions*. London,: Oxford

Le Poidevin, Robin. 2007. *Images of Time: An Essay on Temporal Representation*. Oxford: Oxford University Press

Ladyman, James. 2002. *Understanding Philosophy of Science*. London: Routledge.

Lange, M. 2002. *An Introduction to the Philosophy of Physics*. Oxford: Blackwell Publishing.

Leddington, J. 2009. "Perceptual Presence." *Pacific Philosophical Quarterly* 90: 482–502.

Levin, M. 2000. "Demons, Possibility and Evidence." *Noûs* 34, no. 3: 422–440.

Lewis, David. 1998. "The Problem of Temporary Intrinsics: An Excerpt from on the Plurality of Worlds." In *Metaphysics: The Big Questions*, edited by Van Inwagen, Peter & Zimmerman, Dean, 204–206. London: Routledge.

Lewis, David. 1999. "Rearrangement of Particles: Reply to Lowe." In *Papers in Metaphysics and Epistemology*, 187–195. Cambridge: Cambridge University Press.

Lewis, David & Langton, Rae. 1999. "Defining Intrinsic." In *Papers in Metaphysics and Epistemology*, 116–132. Cambridge: Cambridge University Press.

Locke, Don. 1971. *Memory*. London: MacMillan.

Lockwood, Michael. 1984. "Einstein and Identity Theory." *Analysis* 44, no. 1: 22–25.

Loftus, Elizabeth. 1996. *The Myth of Repressed Memory: False Memories and Allegations of Sexual Abuse*. New York: St. Martin's Griffin.

Logue, Heather. 2011. "The Sceptic and the Naïve Realist." *Philosophical Issues* 21: 268–288.

Logue, Heather. 2015. "Disjunctivism." In *The Oxford Handbook on the Philosophy of Perception*, edited by Mohan Matthen, 198–217. Oxford: Oxford University Press.

Lotto, Beau & Purves, Dale. 2003. *Why We See What We Do: An Empirical Theory of Vision*. Sunderland: Sinauer Associates.

Loux, Michael. 1977. "Ontology." In *The Synoptic Vision: Essays on the Philosophy of Wilfred Sellars*, edited by Delaney, C.F., Loux, M.J., Gutting, G., & Solomon, W.D. London: Notre Dame Press.

Loux, Michael. 2002. *Metaphysics: A Contemporary Introduction*. London: Routledge.

Lowe, E.J. 2003. *A Survey of Metaphysics*. Oxford: Oxford University Press.

Lycan, William. 1996. "Layered Perceptual Representation." *Philosophical Issues* 7: 81–100.

Lycan, William. 2008. "Representational Theories of Consciousness." In *The Stanford Encyclopedia of Philosophy (Fall 2008 Edition)*, edited by Zalta, Edward. http://plato.stanford.edu/archives/fall2008/entries/consciousness-representational.

Mac Cumhaill, Clare. 2013. "Perceiving Immaterial Paths." *Philosophical and Phenomenological Research* 90, no. 3: 687–715.

MacKay, D.S. 1945. "The Illusion of Memory." *The Philosophical Review* 4, no. 104: 297–320.

MacKnik, S.L. & Martinez-Conde, S. 2010. *Sleights of Mind: What the Neuroscience of Magic Reveals about Our Everyday Deceptions*. New York: Henry Holt.

MacPherson, Fiona. 2012a. "The Admissible Contents of Experience: Introduction." In *The Admissible Contents of Experience*, edited by Hawley, Karherine & MacPherson, Fiona. Wiley-Blackwell.

MacPherson, Fiona. 2012b. "Cognitive Penetration of Colour Experience: Rethinking the Issue in Light of an Indirect Mechanism." *Philosophy and Phenomenological Research* 84, no. 1.

Manzotti, R. 2008. "A Process-Oriented View of Qualia." In *The Case for Qualia*, edited by Wright, E., 175–190. London: The MIT Press.

Matthen, Mohan. 2015. "Introduction." In The Oxford Handbook on the Philosophy of Perception, edited by Mohan Matthen, 1–25. Oxford: Oxford University Press.

McCauley, R.N. & Henrich, J. 2006. "Susceptibility to the Müller-Lyer Illusion, Theory-Neutral Observation and the Diachronic Penetrability of the Visual Input System." *Philosophical Psychology* 19, no. 1: 79–101.

McGinn, Colin. 2004. *Mindsight: Image, Dream, Meaning*. London: Harvard University Press.

McTaggart, J.M.E. 1908. "The Unreality of Time." *Mind*, New Series 17, no. 68: 457–474.

McTaggart. J.M.E. 1927. "Chapter 33: Time." In *The Nature of Existence*. Cambridge: Cambridge University Press.

Melia, Joe. 2003. *Modality*. Montreal: McGill-Queens University Press.

Mellor, D. Hugh. 1998. *Real Time II*. London: Routledge.

Miller, Kristie. 2017. "Time Passages." *Journal of Consciousness Studies*, nos. 3–4.

Minkowski, H. 1952. "A Lecture Delivered before the Naturforscher Yersarnmlung (Congress of Natural Philosophers) at Cologne (21st September, 1908)." In *The Principle of Relativity: A Collection of Original Memoirs on the Special and General Theory of Relativity*, edited by Einstein, Albert, Lorentz, H.A., & Minkowski, H., 75–91. New York: Dover.

Molyneaux, B. 2009. "Why Experience Told Me Nothing about Transparency." *Nous* 43, no. 1: 116–136.

Montaigne, Michel. n.d. *Essays of Montaigne, Vol. 9: The Online Library of Liberty*. Accessed February 5th 2013. http://oll.libertyfund.org/?option=com_staticxt&staticfile=show.php%3Ftitle=1750&chapter=91301&layout=html&Itemid=27.

Musgrave, A. 1993. *Common Sense, Science and Scepticism*. Cambridge: Cambridge University Press.

Nanay, Bence. 2012. "The Philosophical Implications of the Perky Experiments: Reply to Hopkins." *Analysis* 72, no. 1: 439–443.

Newman, C.V. & Newman, B. 1974. "Ponzo Illusion in Pictures with and without Suggested Depth." *The American Journal of Psychology* 87, no. 3: 511–516.

Noë, A. 2005. "Real Presence." *Philosophical Topics* 33, no. 1: 235–264.

Noë, A. 2006. "Experience without the Head." In *Perceptual Experience*, edited by Gendler, T.S. & Hawthorne, John, 411–433. Oxford: Oxford University Press.

Nudds, M. 2009. "Recent Work in Perception: Naïve Realism and Its Opponents." *Analysis Reviews* 69, no. 2: 334–346.

Oaklander, L.N. 1995. "Part II Time and Identity." In *Time, Change and Freedom: An Introduction to Metaphysics*, edited by Nathan Oaklander and Quentin Smith, 57–105. London: Routledge.

Oaklander, L.N. 2012. "A-, B- and R-Theories of Time: A Debate." In The Future of the Philosophy of Time, edited by Adrian Bardon, 1–24. New York: Routledge.

O'Callaghan, C. 2009. "Auditory Perception." In *The Stanford Encyclopedia of Philosophy (Summer 2009 Edition)*, edited by Zalta, Edward. http://plato.stanford.edu/archives/sum2009/entries/perception-auditory/.

O'Callaghan, C. 2011. "Lessons from Beyond Vision (Sounds and Audition)." *Philosophical Studies* 153: 143–160.

Pace, M. 2007. "Blurred Vision and the Transparency of Experience." *Pacific Philosophical Quarterly* 88, 3: 328–354.

Paul, L.A. (2010), "Temporal Experience," *Journal of Philosophy*, 107(2): 333–359.

Payne, J.D. & Nadel, L. 2004. "Sleep, Dreams, and Memory Consolidation: The Role of the Stress Hormone Cortisol." *Learning & Memory* 11, no. 6: 671–678.

Perky, C.W. 1910. "An Experimental Study of Imagination." *American Journal of Psychology* 21: 422–452.

Pesoa, L., Thompson, E., & Noë, A. 1998. "Finding Out about Filling-in: A Guide to the Perceptual Completion for Visual Science and the Philosophy of Perception." *Behavioral and Brain Sciences* 21: 723–802.

Plumer, G. 1985. "The Myth of the Specious Present". *Mind*, 94 (373), 19–35.

Power, S.E. 2010. "Complex Experience, Relativity and Abandoning Simultaneity." *Journal of Consciousness Studies* 17, nos. 3–4.

Power, S.E. 2011. "Temporal Illusions: Philosophical Considerations." In *Multidisciplinary Aspects of Time and Time Perception: Lecture Notes in Computer Science 6789*, edited by Vatakis, A., Esposito, A., Giagkou, M., Cummins, F., & Papadelis G. Berlin: Springer.

Power, S.E. 2012. "The Metaphysics of 'the Specious Present'." *Erkenntnis* 77, no. 1: 121–132.

Power, S.E. 2013. "Perceiving External Things and the Time-Lag Argument." [Electronic Version] *European Journal of Philosophy (Early View)*. Accessed September 14th 2012. DOI = 10.1111/j.1468-0378.2010.00436.x0.

Power, S.E. 2016. "Relative and Absolute Temporal Presence." In *Philosophy and Psychology of Time*, edited by Bruno Mölder, Valtteri Arstila, & Peter Øhrstrøm, 67–100. Dordrecht: Springer

Power, S.E. forthcoming. "Against Illusions of Duration." *Chronos*.

Prigatano, G.P. 2010. *The Study of Anosognosia*. Oxford, USA: Oxford University Press.

Prior, Arthur N. 1968. *Papers on Time and Tense*. Oxford: Oxford University Press.

Pritchard, Duncan. 2016. *Epistemology*. Hampshire: Palgrave MacMillan.

Prosser, Simon. 2000. "A New Problem for the A-Theory of Time." *The Philosophical Quarterly* 50, no. 201: 494–498.

Prosser, Simon. 2012. "Why Does Time Seem to Pass?" *Philosophy & Phenomenological Research* 85, no. 1: 92–116.

Reynolds, R.I. 1988. "A Psychological Definition of Illusion." *Philosophical Psychology* 1, no. 2: 217–223.

Reynolds, S.L. 2000. "The Argument from Illusion." *Noûs* 34, no. 4: 604–621.

Robertson, L.C. 2004. *Space, Objects, Minds, and Brains*. Sussex: Psychology Press.

Robinson, H.M. 1974. "The Irrelevance of Intentionality to Perception." *The Philosophical Quarterly* 24, no. 97: 300–315.

Roeckelein, J. 2004. *Imagery in Psychology: A Reference Guide*. Westport: Greenwood Publishing.

Rowlands, Mark. 2010. *The New Science of the Mind: From Extended Mind to Embodied Phenomenology*. London: The MIT Press.

Russell, Bertrand. 1998. "Knowledge by Acquaintance and Knowledge by Description." In *The Problems of Philosophy*, 25–35. London: Oxford University Press.

Sabbagh, Karl. 2009. *Remembering Childhood*. Oxford: Oxford University Press.

Sacks, Oliver. 1996. *An Anthropologist on Mars: Seven Paradoxical Tales*. New York: Vintage.

Sacks, Oliver. 2008. *Musicophilia*. New York: Vintage.

Sacks, Oliver. 2010. "Stereo Sue." In *The Mind's Eye*, 111–143. London: Picador.

Sacks, Oliver. 2012. *Hallucinations*. London: Picador.

Sartre, Jean-Paul. 1986. *The Imaginary: A Phenomenological Psychology of the Imagination*, trans. Webber, J. Oxon: Routledge.

Sellars, Wilfred. 1968. *Science and Metaphysics*. London: Routledge.

Shaw, Julia. 2016. *The Memory Illusion: Remembering, Forgetting, and the Science of False Memory*. London: Random House.

Shoemaker, Sydney. 1993. "Time without Change." In *Philosophy of Time*, edited by Le Poidevin, Robin & MacBeath, Murray. Oxford: Oxford University Press.

Sider, Ted. 2003. *Four-Dimensionalism: An Ontology of Persistence and Time*. London: Oxford University Press.

Siegel, S. 2006. "Subject and Object in the Contents of Visual Experience." *Philosophical Review* 115, no. 3.

Siegel, S. 2011. "The Contents of Perception." In *The Stanford Encyclopedia of Philosophy (Winter 2011 Edition)*, edited by Zalta, Edward. http://plato.stanford.edu/archives/win2011/entries/perception-contents.

Siegel, Susanna, & Silins, Nicholas. 2015. "The Epistemology of Perception." In The Oxford Handbook on the Philosophy of Perception, edited by Mohan Matthen, 781–812. Oxford: Oxford University Press.

Skow, B. 2012. "Why Does Time Pass?" *Nous* 46, no. 2: 223–242.

Smart, J.M.C. 2001. "The Space-Time World." In *Metaphysics: Contemporary Readings*, edited by Loux, Michael. London: Routledge.

Smith, A.D. 2010. "Disjunctivism and Illusion." *Philosophy and Phenomenological Research* 80, no. 22: 384–410.

Smith, B.C. unpublished manuscript. "Rehabilitating Representation." *n/a, n/a.*

Smith, Q. 1995. *Time, Change and Freedom: An Introduction to Metaphysics*. London: Routledge.

Soteriou, M. 2010. "The Disjunctive Theory of Perception." In *The Stanford Encyclopedia of Philosophy (Winter 2010 Edition)*, edited by Zalta, Edward. http://plato.stanford.edu/archives/win2010/entries/perception-disjunctive.

Torrengo, G. 2017. "Feeling the Passing of Time." *Journal of Philosophy* 114, no. 4.

Tye, Michael. 1975. "The Adverbial Theory: A Defence of Sellars against Jackson." *Metaphilosophy* 6: 136–143.

Varela, F. 1999. "The Specious Present: A Neurophenomenology of Time Consciousness." In *Naturalizing Phenomenology*, edited by Petitot, J., Varela, F.J., Pacoud, B., & Roy, J. Stanford: Stanford University Press.

Wackermann, J. 2012. "Determinants of Filled/Empty Optical Illusion: Differential Effects of Patterning." *Acta Neurobiologiae Experimentalis* 72, no. 1: 89–94.

West, L.J. 2011. "Hallucinations." *Encyclopedia Britannica Online*. www.britannica.com/EBchecked/topic/252916/hallucination.

Wittgenstein, Ludwig. 1974. *Tractatus Logico-Philosophicus*, trans. Pears, D.F. & McGuinness, B.F. Oxon: Routledge.

Wittmann, Marc. 2011. "Moments in Time." *Frontiers in Integrative Neuroscience* 5, no. 66. DOI = 10.3389/fnint.2011.00066.

Zahavi, D. 2007. "Perception of Duration Presupposes Duration of Perception—or Does It? Husserl and Dainton on Time." *International Journal of Philosophical Studies* 15, no. 3: 453–471.

Zahavi, D. 2008. "Internalism, Externalism and Transcendental Idealism." *Synthese* 160: 355–374.

Zaidi, Q., Ennis, R., Cao, D., & Lee, B. 2012. "Neural Locus of Color Afterimages." *Current Biology* 22, no. 3: 220–224.

Index

Alston, W. 67–68
anosognosia 89; and apparent limitations of experience 90; and change blindness 94; definition 92
Arstila, V. 70, 111

Bardon, A. 173, 182
Bergson, H. 29, 187
Blackmore, S. 95
Breckon, T. 62–63
Brentano, F. 58
Broad, C. 24
Butterfield, J. 40, 133

Callender, C. 40, 126–127
Chalmers, D. 70
change blindness, *see* anosognosia
change: perceptual experience of 169; real change 22; and the 'specious' present 169–172; tenseless change 24
Changizi, M. 137, 142, 161
Clark, A. 3, 70
common features of illusion and hallucination 73
conflationary errors 78
contingent things 4
Crane, T. 57, 67, 69, 70, 82
cubist vision 155–157

Dainton, B. 30, 187, 193, 197
Dancy, J. 74
De Warren, N. 194
Debus, D. 213
Deng, N. 36, 181
Dennett, D. 173, 182
depth: apparent 141–145; and geometric-optical illusions 161–163; hidden 145–147; and mirages 152–163; and the time-lag argument 147–149
duration: anosognosia of 160, 174–176, 215–217; experience of 168–171; extensionalism and 197–198; hidden duration 132–135, 225, 234–237; of perception 128–129; relative duration 31–34; retention theory and 194; *see also* durational present

embedded mind, *see* extended mind hypothesis
erroneous experiences vs erroneous beliefs 77
eternalism, definition of 25
experience: amodality of 61–63; erroneous 5; extended and embedded 46; hidden 58; imaginative experience 43–44, 58, 62–65, 168, 171, 205–206; justificatory role of 4; memory experience 40–44, 62, 70, 129, 167–169, 171, 192, 200, 206–211, 213–214; non-perceptual experience 40; perceptual experience 4; vs perception 4; reliability (vs trustworthiness) of 1; representationalism and 64–68; structure of 46–48; relations in 53–59; subjective activity of 59; varieties of 40
extended mind hypothesis (EMH), 48–50
extensionalism 197–199
externalism (theories of knowledge), 2
externality 114–115

Fish, W. 82
Fisher, R. 62–63
frames of reference 31

Gibson, J. 141–142
growing block theory 35, 37
Grush, R. 66, 70, 182, 192, 201

hallucination: and complex illusions 85–86; definition 80; eliminative theories of 212–218; evidence of 203–205; imaginative theories of 205–206; and internal things 83; memory theories of 206–211; and mirages 83–84; object of 82–83; and sortal illusions 87–88; temporal hallucination 172; *see also* illusion
Houts, R. 117–118
Husserl, E. 41, 66, 192–194

illusion: complex illusion 84; definition 77; geometric-optical illusion (GOI), 72–73; non-perceptual illusion 43; sortal illusions 87; of totality 109–110
intervals, relativistic 34

Jackson, F. 58
James, W. 180

Kahneman, D. 92
Kalderon, M. 78
knowledge of the actual world 3
Kölers, P. 41–43, 162

Langton, R. 44, 54
Le Poidevin, R. 40, 126, 172, 175, 208
Leddington, J. 57
Lewis, D, 44, 54
Loftus, E. 234
Logue, H. 45, 56–57, 67
Loux, M. 86
Lowe, E. 22, 26, 27
Lycan, W. 66

Mac Cumhaill, C. 145
MacKay, D. 41
MacPherson, F. 43, 64–65
magic 72–73, 102–103
Manzotti, R. 215
Matthen, M. 4–5
McTaggart, J. 23–24
Mellor, D. 27, 33, 126, 137, 236
mental events, the non-spatiality of 100–102
Miller, K. 180, 181
Minkowski, H. 34, 51
Montaigne, M. 3–4
moving spotlight theory 23–24

naïve realism 44
Nanay, B. 43
Noë, A. 62
Nudds, M. 45

O'Callaghan, C. 95–96
objective activity 129–131
ontological commitment 7–10

Paul, L. 182
perception: and the philosophy of time 132–135; the perceptual relation 56–59; the process of 129–135
Perky effect 43–44
present: durational present 127, 137; experience of 40–41; and external things 124; perceiving the 116–117, 125–128; punctual present 127, 170; 'specious' present 41, 169–171
presentism, definition of 25
Pritchard, D. 2, 14
property theory: and apparent objects 84; property-bearers and hallucination 86

reality in time 25
relativity 27; exceptions in physics 33; in physics 28–29; of space and time 31–32; relative and absolute 187–189; and subjectivity 29
retention theory 192–197
retentionalism, *see* retention theory
Rowlands, M. 48–49
Russell, B. 41, 23

Sacks, O. 89, 101, 144, 203, 223
Siegel, S. 5
Silins, N. 5
simultaneity, relativity of 22, 29–32, 50, 187–188
Soteriou, M. 82
spatial relations: and the past 117–122; phenomenology of perceived spaced 114–116 (*see also* depth, apparent and the philosophy of time 122–124); and relativity 118–119

St. Augustine 117–118
subjective: activity 60–62, 131–132; lag 215–225; relativity and subjectivity 29–30
succession, *see* temporal order

temporal experience: models of 191–199; perceptual experiences of time 168–172
temporal illusion and temporal anosognosia 173–175
temporal order: anosognosia of 185–187; extensionalism and 197; illusions of 181–185; (as succession) in apparent motion 173; (as succession) perception of 127; (as succession) of feelings 201; (as succession) relative temporal order 31–33, 190
temporal passage: definition 24–25; experience of 179–181; relative properties and 189–190
temporal structure 34; static structure 34; dynamic structure 34
tense theory, definition of 26
tenseless theory, definition of 26
Tye, M. 58

Wackermann, J. 72
Wittgenstein, L. 155
WYSIATI ('What You See Is All There Is') 92

For Product Safety Concerns and Information please contact our EU
representative GPSR@taylorandfrancis.com
Taylor & Francis Verlag GmbH, Kaufingerstraße 24, 80331 München, Germany

www.ingramcontent.com/pod-product-compliance
Ingram Content Group UK Ltd.
Pitfield, Milton Keynes, MK11 3LW, UK
UKHW022001180926
475701UK00003B/132
* 9 7 8 0 3 6 7 6 6 7 0 8 5 *